MÉMOIRES DE LA SMF 100

ON SUMS OF SIXTEEN BIQUADRATES

Jean-Marc Deshouillers

Koichi Kawada

Trevor D. Wooley

Société Mathématique de France 2005
Publié avec le concours du Centre National de la Recherche Scientifique

J.-M. Deshouillers

Statistique Mathématique et Applications (EA 2961),
Université Victor Segalen Bordeaux 2, 33076 Bordeaux, France &
A2X, Université Bordeaux 1 and CNRS (UMR 5465), 33405 Talence, France.
E-mail : jean-marc.deshouillers@math.u-bordeaux1.fr

K. Kawada

Department of Mathematics, Faculty of Education, Iwate University,
Morioka, 020-8550 Japan.
E-mail : kawada@iwate-u.ac.jp

T.D. Wooley

Department of Mathematics, University of Michigan,
East Hall, 525 East University Avenue, Ann Arbor, Michigan 48109-1109, U.S.A..
E-mail : wooley@umich.edu
Url : http://www.math.lsa.umich.edu/~wooley

2000 *Mathematics Subject Classification*. — Primary 11P05, 11P55; Secondary 11D45, 11D85, 11L15, 11N56.

Key words and phrases. — Waring's problem, circle method, Weyl sums, multiplicative functions, Diophantine equations.

J.-M. D.: Supported by the universities Bordeaux 1, Victor Segalen Bordeaux 2 and CNRS (UMR 5465).
T.D. W.: Packard Fellow, and supported in part by NSF grant DMS-9970440.

ON SUMS OF SIXTEEN BIQUADRATES

Jean-Marc Deshouillers, Koichi Kawada, Trevor D. Wooley

Abstract. — By 1939 it was known that 13792 cannot be expressed as a sum of sixteen biquadrates (folklore), that there exist infinitely many natural numbers which cannot be written as sums of fifteen biquadrates (Kempner) and that every sufficiently large integer is a sum of sixteen biquadrates (Davenport).

In this memoir it is shown that every integer larger than 10^{216} and not divisible by 16 can be represented as a sum of sixteen biquadrates. Combined with a numerical study by Deshouillers, Hennecart and Landreau, this result implies that every integer larger than 13792 is a sum of sixteen biquadrates.

Résumé (Sur les Sommes de Seize Bicarrés). — En 1939, on savait que 13792 ne peut pas être représenté comme somme de seize bicarrés (folklore), qu'il existe une infinité d'entiers qui ne peuvent pas être écrits comme sommes de quinze bicarrés (Kempner) et que tout entier assez grand est somme de seize bicarrés (Davenport).

Dans ce mémoire, on montre que tout entier supérieur à 10^{216} et non divisible par 16 peut s'exprimer comme somme de seize bicarrés. Combiné à une étude numérique menée par Deshouillers, Hennecart et Landreau, ce résultat implique que tout entier supérieur à 13792 est somme de seize bicarrés.

CONTENTS

1. Introduction 1
2. Outline of the proof of Theorem 2 7
3. The cardinality of the sets $\mathcal{M}_\eta(X)$ 13
4. An auxiliary singular integral 27
5. Estimates for complete exponential sums 31
6. The singular series 43
7. The major arc contribution 53
8. An explicit version of Weyl's inequality 59
9. An auxiliary bound for the divisor function 67
10. An investigation of certain congruences 85
11. Mean value estimates 99
12. Appendix : Sums of nineteen biquadrates 105
Bibliography 119

CHAPTER 1

INTRODUCTION

The investigation of sums of biquadrates occupies a distinguished position in additive number theory, largely on account of the relative success with which the basic problems of Waring-type have been addressed. Although progress on such problems was dominated for the greater part of the 20th century by advances in technology at the heart of the Hardy-Littlewood method, the older ideas involving the use of polynomial identities have recently resurfaced in work of Kawada and Wooley [15], though now interwoven with the analytic machinery of the circle method itself. The primary goal of this paper is to apply this new circle of ideas to obtain an explicit analysis of sums of sixteen biquadrates, and, moreover, one suitable for determining the largest integer not represented in such a manner. A separate paper [12] reports on computations of Deshouillers, Hennecart and Landreau which complement the main conclusion of this memoir, and as will shortly become apparent, the union of these results leads to the following definitive statement concerning sums of sixteen biquadrates.

THEOREM 1. — *Every integer exceeding* 13792 *can be written as a sum of at most 16 biquadrates.*

Although we avoid a detailed historical account of the various contributions to Waring's problem for biquadrates, our subsequent discussion will be clarified by a sketchy overview of such matters (we refer the reader to the survey [11] for a more comprehensive account). For the sake of concision, we refer to a number n as being a B_s (number) when n can be written as a sum of at most s biquadrates. In accordance with the familiar notation in Waring's problem, we then denote by $g(4)$ the least integer s with the property that every natural number is a B_s, and we denote by $G(4)$ the least natural number s such that every sufficiently large number is a B_s. The problem central to this paper has as its origin the assertion made by Waring in 1770 to the effect that $g(4) = 19$. This conjecture was in large part resolved by Hardy and

Littlewood [**13**], who established by means of their newly devised circle method that $G(4) \leqslant 19$. Indeed, the work of Hardy and Littlewood shows that one may compute an explicit constant C with the property that every number exceeding C is a B_{19}. Although a computational check of the integers of size at most C would determine whether or not $g(4)$ is equal to 19, the astronomical size of this constant C entirely precluded any such attempt to resolve this problem. While for other exponents k, advances in the circle method rapidly wrought an effective determination of the value of $g(k)$, it was only in the late 1980's that, with new ideas and substantial effort, it became possible to reduce the value of C to a size within the grasp of existing supercomputers. Thus Balasubramanian, Deshouillers and Dress at last announced a proof of $g(4) = 19$ in [**3**], [**4**]. A complete proof of the result can be found in the series of papers [**7**], [**8**], [**9**] and [**10**].

While it has only recently been established that every natural number is a B_{19}, as Waring had claimed, it has been known for many years that $G(4)$ is less than 19. Indeed, Davenport [**5**] had shown by 1939 that $G(4) = 16$, so that with only finitely many exceptions, all natural numbers are B_{16}. We recall at this point that the lower bound $G(4) \geqslant 16$ is immediate from the observation that $31 \cdot 16^m$ is not a B_{15} for any non-negative integer m. As announced in [**11**], by combining the work of Balasubramanian, Deshouillers and Dress with the central idea of the recent memoir [**15**] of Kawada and Wooley, it is now possible to determine all numbers that are not B_{16}. The object of this treatise is the detailed proof of the following result.

THEOREM 2. — *Every integer exceeding 10^{216} that is not divisible by 16 can be written as the sum of 16 biquadrates.*

A companion paper of Deshouillers, Hennecart and Landreau [**12**] shows that all natural numbers not exceeding 10^{245} are B_{16}, with the exception of precisely 96 numbers, the largest of which is 13792. In view of the latter conclusion, Theorem 1 follows from Theorem 2 by noting that integers exceeding 10^{216} divisible by 16 are harmless. For if $N > 10^{216}$ and $16|N$, then there exist natural numbers m and n with the property that $N = 16^m n$, and either $n > 10^{216}$ and $16 \nmid n$, or else $10^{216}/16 < n \leqslant 10^{216}$. In the former case, Theorem 2 shows that n is a B_{16}, and in the latter case the above cited conclusion of Deshouillers, Hennecart and Landreau [**12**] shows that n is a B_{16}. Thus, in either case, it is evident that $N = (2^m)^4 n$ is a B_{16}.

We remark that Deshouillers, Hennecart and Landreau [**12**] have determined in addition the 31 numbers that are not B_{17} (the largest of which is 1248), and also the 7 numbers that are not B_{18}, these being simply described as the integers $80k - 1$ for $1 \leqslant k \leqslant 7$. We refer the reader to the aforementioned paper [**12**] for a complete list of the exceptional numbers which are not B_{16}, and those which are not B_{17} (this information may also be found in the survey [**11**]).

We next provide a brief overview of our basic strategy, deferring to section 2 a more detailed discussion of our plan of attack on the proof of Theorem 2. We employ the Hardy–Littlewood method, aiming to exploit the polynomial identity

$$x^4 + y^4 + (x+y)^4 = 2(x^2 + xy + y^2)^2 \tag{1.1}$$

that was the key innovation of Kawada and Wooley [15]. In order to efficiently exploit the relation (1.1), we introduce the set $\mathcal{M}$, which we define by

$$\mathcal{M} = \{ m \in \mathbb{N} : m = x^2 + xy + y^2 \text{ for some } x, y \in \mathbb{Z} \text{ with } xy(x+y) \neq 0 \}. \tag{1.2}$$

In view of (1.1), for each $m \in \mathcal{M}$ one finds that $2m^2$ is a sum of 3 biquadrates. Thus one is led to consider the number, $\mathcal{Z}(X)$, of solutions of the equation

$$2m_1^2 + u_1^4 + u_2^4 = 2m_2^2 + u_3^4 + u_4^4,$$

with $m_1, m_2 \in \mathcal{M} \cap [1, X^{1/2}]$ and $1 \leqslant u_i \leqslant X^{1/4}$ ($1 \leqslant i \leqslant 4$). By employing a modified divisor function estimate to determine the number of solutions of the latter equation with $u_1^4 + u_2^4 \neq u_3^4 + u_4^4$, and an immediate counting argument when $u_1^4 + u_2^4 = u_3^4 + u_4^4$, one derives the efficient upper bound $\mathcal{Z}(X) \ll X(\log X)^\varepsilon$ without any undue effort (see the proof of Theorem 1 in Kawada and Wooley [15, §2], and also the related discussion of Lemma 3.4 of [15]).

In order to establish that a given large number N is a B_{16}, the most obvious strategy suggested by the above discussion is that of considering representations of N in the form

$$N = 2m_1^2 + 2m_2^2 + x_1^4 + \cdots + x_{10}^4, \tag{1.3}$$

with $m_1, m_2 \in \mathcal{M}$ and $x_j \in \mathbb{N}$ ($1 \leqslant j \leqslant 10$). It is now apparent that whenever N admits a representation of the shape (1.3), then N may be written as the sum of 16 biquadrates. Unfortunately, since a biquadrate is congruent to 0 or 1 modulo 16 according to whether it is even or odd, one finds from (1.1) that whenever $m \in \mathcal{M}$, the expression $2m^2$ is necessarily congruent to 0 or 2 modulo 16. Thus, whereas an unrestricted sum of three biquadrates is congruent to 0, 1, 2 or 3 modulo 16, our surrogate $2m^2$ is restricted to the classes 0 and 2 modulo 16. It follows that whether or not the integer N is a B_{16}, it fails to possess a representation in the shape (1.3) whenever $N \equiv 15 \pmod{16}$, and thus our initial strategy is doomed to failure. Nonetheless, by making use of the tools developed within this memoir, the authors have employed this approach to establish that whenever $N \geqslant 10^{156}$, and N is not congruent to 0 or 15 modulo 16, then N can be written in the shape (1.3), and hence is a B_{16}. We omit the details of such an argument in the interest of saving space.

As is apparent from the deliberations of the previous paragraph, one may recover the missing congruence class 15 modulo 16 by considering instead representations of N in the form

$$N = 2m^2 + x_1^4 + \cdots + x_{13}^4, \tag{1.4}$$

with $m \in \mathcal{M}$ and $x_j \in \mathbb{N}$ ($1 \leqslant j \leqslant 13$). By combining the mean value estimate discussed above with an explicit version of Hua's inequality provided by Deshouillers and Dress [9], a careful application of the Hardy-Littlewood method would establish that whenever $N \geqslant 10^{300}$ and $16 \nmid N$, then N is a B_{16}. Unfortunately, even anticipated advances in computational technology would seem insufficient to permit the methods of Deshouillers, Hennecart and Landreau [12] to check that all numbers not exceeding 10^{300} are B_{16}, with the above-mentioned exceptions. We are therefore forced in our proof of Theorem 2 to introduce a further new idea.

Motivated by the identity (1.1), and the similar identity
$$x^2 + y^2 + (x+y)^2 = 2(x^2 + xy + y^2),$$
we obtain from the relation
$$(w+x)^4 + (w-x)^4 = 2w^4 + 12w^2x^2 + 2x^4$$
the new identity
$$\begin{aligned}(w+x)^4 + (w-x)^4 &+ (w+y)^4 + (w-y)^4 + (w+x+y)^4 + (w-x-y)^4 \\ &= 4(x^2 + xy + y^2)^2 + 24(x^2 + xy + y^2)w^2 + 6w^4 \\ &= 4(x^2 + xy + y^2 + 3w^2)^2 - 30w^4.\end{aligned} \quad (1.5)$$

The use of (1.1) in the representations (1.3) and (1.4) might reasonably be regarded as effectively replacing three biquadrates by a square. The use of the identity (1.5), meanwhile, effectively replaces six biquadrates by a square and a biquadrate, which in applications amounts to trading five biquadrates for a square. While the latter exchange is clearly less efficient than the former so far as consequent mean value estimates are concerned (see Lemmata 2.4 and 2.5 below), in compensation one finds that the six biquadrates on the left hand side of (1.5) may be simultaneously odd. Moreover, despite the relative inefficiency of the identity (1.5) as compared to (1.1), one may nonetheless recover a mean value estimate associated with only 14 biquadrates of essentially the same strength as that available from Hua's inequality for 16 biquadrates (compare Theorem 4 of Deshouillers and Dress [9] with Lemma 2.5 below). Thus it transpires that the new identity (1.5) is crucial to the success of this paper.

In order to establish that a given large integer N is a B_{16}, therefore, the strategy which we adopt in this memoir is to consider representations of N in the form
$$N = 2m_1^2 + 4m_2^2 + 24m_2w^2 + 6w^4 + x_1^4 + \cdots + x_7^4,$$
with $m_1, m_2 \in \mathcal{M}$ and $w, x_j \in \mathbb{N}$ ($1 \leqslant j \leqslant 7$). In view of the identities (1.1) and (1.5), it follows that whenever N can be written in the latter form, then N is necessarily a sum of 16 biquadrates. A discussion of the details associated with putting this strategy into practice may be found in §2 below, wherein an outline of the proof of Theorem 2 is also provided.

CHAPTER 1. INTRODUCTION

Equipped now with the powerful new weapons developed for our attack on sums of 16 biquadrates, it is difficult to resist the temptation to return to the topic of sums of 19 biquadrates. As mentioned earlier, Deshouillers and Dress showed first in [**9**] that every number exceeding 10^{367} is a B_{19}, and subsequently in [**10**], with heavy use of powerful computers, that every number up to 10^{448} is a B_{19}. In an appendix to this paper we apply our idea based on the use of the identity (1.1) in order to substantially reduce the computations necessary to establish that $g(4) = 19$. We show in fact that every number N exceeding 10^{146} can be represented in the form

$$N = 2m_1^2 + 2m_2^2 + x_1^4 + \cdots + x_{13}^4,$$

with m_1, $m_2 \in \mathcal{M}$ and $x_j \in \mathbb{N}$ ($1 \leqslant j \leqslant 13$), whence by (1.1) it follows that every such N is a B_{19}. The computational verification that every number up to 10^{147} is a B_{19} can be completed even on a modest personal computer within a few minutes.

We finish our opening remarks with a few comments concerning the extent to which numerical computations underlie the main conclusions of this memoir. While the contribution of Deshouillers, Hennecart and Landreau [**12**] to the proof of Theorem 1 is necessarily heavy in its use of powerful computers, we have expended considerable effort in our proof of Theorem 2 in avoiding serious computations, either explicit, or implicit in results cited from the literature. Indeed, a moderately energetic reader equipped only with a hand-held calculator should encounter no difficulties in verifying the computations involved in this analytic argument. A more cavalier approach to the use of computational results, and especially those to be found in the literature, would improve the conclusion of Theorem 2 somewhat. In particular, our Lemma 3.2 below could be replaced by Theorem 5 of Rosser and Schoenfeld [**19**], our Lemma 3.3 could be replaced by Theorem 5.3 of McCurley [**17**], and our Lemma 2.3 could be improved by the use of the numerical estimates for infinite products recorded on p.295 of Deshouillers [**7**]. One might also make use of numerical integration to evaluate the singular integral, rather than using Lemma 4.2 below. Incorporating such computational refinements into our basic argument, one may show that every integer exceeding 10^{196}, and not divisible by 16, can be written as the sum of 16 biquadrates, thereby improving the corresponding bound of Theorem 2 by a factor of 10^{20}.

Throughout this paper, we write $e(z)$ for $e^{2\pi i z}$. Also, we denote the largest integer not exceeding x by $[x]$, and we write $\lceil x \rceil$ for the least integer y with $y \geqslant x$. Also, the letter p will always be reserved to denote a prime number.

CHAPTER 2

OUTLINE OF THE PROOF OF THEOREM 2

As promised in the introduction, we now outline the proof of Theorem 2 in some detail. We begin by defining the exponential sums and arc dissections at the heart of our application of the Hardy-Littlewood method. Here, in order more easily to exploit previous work on the problem, we make heavy use of the notation of [6] and [9]. Thus, when P is a natural number and $\varepsilon \in \{0,1\}$, we define the exponential sum $S_\varepsilon(\alpha)$ by

$$S_\varepsilon(\alpha) = \sum_{P-\varepsilon/2 < x \leqslant 2P-\varepsilon/2} e((2x+\varepsilon)^4 \alpha). \tag{2.1}$$

We define the unit interval $\mathfrak{U}$ by

$$\mathfrak{U} = [975P^{-3},\ 1+975P^{-3}].$$

Also, when $a \in \mathbb{Z}$ and $q \in \mathbb{N}$, we define the set

$$\mathfrak{M}(q,a) = \{\alpha \in \mathfrak{U}\ :\ |q\alpha - a| \leqslant 975P^{-3}\}, \tag{2.2}$$

and then define the set of major arcs $\mathfrak{M}$ to be the union of the intervals $\mathfrak{M}(q,a)$ with $0 \leqslant a \leqslant q \leqslant P^{1/2}$ and $(a,q)=1$. Note that the intervals occurring in the latter union are disjoint whenever P is large enough, and such is certainly the case for $P \geqslant 100$. Finally, we define the minor arcs $\mathfrak{m}$ by $\mathfrak{m} = \mathfrak{U} \setminus \mathfrak{M}$.

We next turn our attention to the important set $\mathcal{M}$ defined in (1.2), extending our earlier notation by writing, for $\eta \in \{0,1\}$,

$$\mathcal{M}_\eta = \{m \in \mathcal{M}\ :\ m \equiv \eta \pmod{2}\ \}$$

and

$$\mathcal{M}_\eta(X) = \mathcal{M}_\eta \cap [1, X].$$

Notice that whenever $m \in \mathcal{M}_0$, one necessarily has $4|m$, and so it follows that when $m \in \mathcal{M}_\eta$ with $\eta \in \{0,1\}$, then

$$2m^2 \equiv 2\eta \pmod{16}. \tag{2.3}$$

Also, when $m \in \mathcal{M}_0$ and $\zeta \in \{0, 1\}$, we find that for any integer w, one has
$$4m^2 + 24m(2w + \zeta)^2 + 6(2w + \zeta)^4 \equiv 6\zeta \pmod{16}. \tag{2.4}$$

Since it is possible, without excessive inconvenience, to restrict all the biquadrates employed in our representation to be non-zero, we define our generating functions in such a way as to make this possible. Accordingly, when $m \in \mathcal{M}_0$, we consider the unique integers x and y satisfying $m = x^2 + xy + y^2$ with $x \geqslant y$ and $x + y$ largest amongst all the latter representations, and then put
$$\mathcal{W}(m) = \{|x|, |y|, |x + y|\}. \tag{2.5}$$

On recalling the identity (1.5), we may express the left hand side of (2.4) as a sum of six biquadrates. When $w > 0$ and $2w + \zeta \notin \mathcal{W}(m)$, moreover, it is apparent that all six biquadrates occurring in the latter expression are non-zero. When P is a natural number, and $\eta, \zeta \in \{0, 1\}$, we define the exponential sums
$$F_\eta(\alpha) = \sum_{m \in \mathcal{M}_\eta(P^2)} e(2m^2 \alpha), \tag{2.6}$$
and
$$D_\zeta(\alpha) = \sum_{\substack{m \in \mathcal{M}_0(3P^2/7)}} \sum_{\substack{1 \leqslant w < P/6 \\ 2w + \zeta \notin \mathcal{W}(m)}} e((4m^2 + 24m(2w+\zeta)^2 + 6(2w+\zeta)^4)\alpha). \tag{2.7}$$

Motivated by the need to satisfy relevant congruence conditions, when $N \equiv r \pmod{16}$ with $1 \leqslant r \leqslant 15$, we define the integers η, ζ and t by
$$\begin{cases} \eta = 0, \zeta = 0 \text{ and } t = r & \text{for } 1 \leqslant r \leqslant 2, \\ \eta = 1, \zeta = 0 \text{ and } t = r - 2 & \text{for } 3 \leqslant r \leqslant 8, \\ \eta = 1, \zeta = 1 \text{ and } t = r - 8 & \text{for } 9 \leqslant r \leqslant 15. \end{cases} \tag{2.8}$$

We note that in any case, our choices for η, ζ and t ensure that
$$1 \leqslant t \leqslant 7 \quad \text{and} \quad N - 2\eta - 6\zeta \equiv t \pmod{16}. \tag{2.9}$$

Also, when N is a natural number and ν is a positive real number, we define the positive numbers $P_0 = P_0(N, \nu)$ and $P = P(N, \nu)$ by means of the relations
$$N = 16\nu P_0^4 \quad \text{and} \quad P = [P_0]. \tag{2.10}$$

Equipped with the above notation, we denote by $R(N) = R(N, \nu)$ the number of representations of the natural number N in the form
$$N = 2m_1^2 + 4m_2^2 + 24m_2(2w+\zeta)^2 + 6(2w+\zeta)^4 + \sum_{j=1}^{7-t}(2x_j)^4 + \sum_{l=1}^{t}(2y_l + 1)^4,$$
subject to
$$m_1 \in \mathcal{M}_\eta(P^2), \quad m_2 \in \mathcal{M}_0(3P^2/7), \quad 1 \leqslant w < P/6, \quad 2w + \zeta \notin \mathcal{W}(m_2),$$
$$P < x_j \leqslant 2P \ (1 \leqslant j \leqslant 7 - t) \quad \text{and} \quad P \leqslant y_l < 2P \ (1 \leqslant l \leqslant t).$$

Thus, in view of the identities (1.1) and (1.5), together with the definitions of the sets $\mathcal{M}_\eta(X)$ and $\mathcal{W}(m)$, it follows that whenever $R(N) > 0$, then N can be written as a sum of 16 biquadrates.

Next, when $\mathfrak{L} \subseteq \mathfrak{U}$, we define $R(N; \mathfrak{L}) = R(N, \nu; \mathfrak{L})$ by

$$R(N; \mathfrak{L}) = \int_{\mathfrak{L}} F_\eta(\alpha) D_\zeta(\alpha) S_0(\alpha)^{7-t} S_1(\alpha)^t e(-N\alpha) d\alpha, \tag{2.11}$$

and observe that by orthogonality, one has

$$R(N) = R(N; \mathfrak{U}) = R(N; \mathfrak{M}) + R(N; \mathfrak{m}). \tag{2.12}$$

We emphasise here that the quantities P, $R(N;\mathfrak{M})$ and $R(N;\mathfrak{m})$ should be regarded as functions of both N and ν. By evaluating $R(N;\mathfrak{M})$ and $R(N;\mathfrak{m})$, we aim to show that when $N \geqslant 10^{216}$, then there exists a positive number ν with the property that $R(N) > 0$. Theorem 2 evidently follows immediately from the latter conclusion.

We estimate the contribution of $R(N;\mathfrak{M})$ by making use of the tools supplied in [6] and [9]. After preparing some auxiliary estimates for exponential sums in §5, we estimate the singular series in §6. Combining the latter with an estimate for the singular integral obtained in §4, we derive in §7 the following lower bound for $R(N;\mathfrak{M})$.

LEMMA 2.1. — *Suppose that $P \geqslant 10^{50}$, and write*

$$M_\eta = \operatorname{card}(\mathcal{M}_\eta(P^2)) \quad \text{and} \quad \widetilde{M}_0 = \operatorname{card}(\mathcal{M}_0(3P^2/7)). \tag{2.13}$$

Then there exists a positive real number ν, with $\nu < 64$, such that

$$R(N;\mathfrak{M}) > 0.00021 M_\eta \widetilde{M}_0 P^4.$$

In order to deduce a satisfactory lower bound for $R(N;\mathfrak{M})$, we require explicit lower bounds for M_η and $\widetilde{M}_0$, and these we establish in §3. We summarise these bounds in the following lemma.

LEMMA 2.2. — *When $X \geqslant 10^{60}$, one has*

$$\operatorname{card}(\mathcal{M}_0(X)) > 0.0508 \frac{X}{\sqrt{\log X}} \quad \text{and} \quad \operatorname{card}(\mathcal{M}_1(X)) > 0.1524 \frac{X}{\sqrt{\log X}}.$$

We estimate the contribution of $R(N;\mathfrak{m})$ by combining an explicit version of Weyl's inequality with certain mean value estimates based on the polynomial identities (1.1) and (1.5). So far as Weyl's inequality is concerned, we note that Deshouillers [7] has made use of an idea of Balasubramanian [2] in order to provide an explicit bound valid for $P \geqslant 10^{80}$. Since our application demands the use of smaller values of P, in §8 we modify the argument described in [7] so as to establish the following conclusion.

LEMMA 2.3. — *Suppose that $\varepsilon \in \{0,1\}$. Then whenever $P \geqslant 10^{30}$ one has*

$$\sup_{\alpha \in \mathfrak{m}} |S_\varepsilon(\alpha)| \leqslant 77 P^{0.884} (\log P)^{0.25},$$

and when $P \geqslant 10^{53}$ one has the sharper bound

$$\sup_{\alpha \in \mathfrak{m}} |S_\varepsilon(\alpha)| \leqslant 16.7 P^{0.884} (\log P)^{0.25}.$$

We turn our attention next to mean value estimates employed on the minor arcs, and it is at this point that we profit handsomely from the identities (1.1) and (1.5). By making use of a well-known property of the divisor function, one swiftly deduces that for each ε, η, ζ, one has upper bounds of the shape

$$\int_0^1 |F_\eta(\alpha)^2 S_\varepsilon(\alpha)^4| d\alpha \leqslant A_1 P^4 (\log P)^{C_1}$$

and

$$\int_0^1 |D_\zeta(\alpha)^2 S_\varepsilon(\alpha)^2| d\alpha \leqslant A_2 P^4 (\log P)^{C_2},$$

for suitable constants A_i, C_i ($i = 1, 2$). For example, the general methods of van der Corput, and of Wolke, would be adequate to prove such an inequality. However, such general methods yield excessively large values of C_i, and in such circumstances the quality of these bounds becomes completely ineffective for the purpose at hand. We therefore adopt a strategy modelled on the argument described in Deshouillers and Dress [8] and [9], deriving an auxiliary upper bound for the divisor function in §9 (see, in particular, the conclusion of Lemma 9.10), bounding the number of solutions of congruences employed within the argument in §10, and finally establishing the crucial mean value estimates in §11. Thus we obtain the following two lemmata.

LEMMA 2.4. — *Suppose that $P \geqslant 10^{25}$ and $\varepsilon, \eta \in \{0, 1\}$. Then one has*

$$\int_0^1 |F_\eta(\alpha)^2 S_\varepsilon(\alpha)^4| d\alpha < 60 M_\eta P^2 (\log P)^4 + 500 P^4 (\log P)^3 + 15360 P^4 (\log P)^2,$$

where M_η is defined as in (2.13).

LEMMA 2.5. — *Suppose that $P \geqslant 10^{50}$ and $\zeta \in \{0, 1\}$. Then one has*

$$\int_0^1 |D_\zeta(\alpha)^2 S_1(\alpha)^2| d\alpha < 10 \widetilde{M_0} P^2 (\log P)^4 + 45 P^4 (\log P)^3,$$

where $\widetilde{M_0}$ is defined as in (2.13).

Equipped with the previous five lemmata, the proof of Theorem 2 is swiftly overwhelmed via a straightforward computation, as we now demonstrate. We suppose in what follows that $P \geqslant 10^{53}$, and begin by observing that as a consequence of Lemma 2.2, one has

$$M_\eta > \frac{0.0508 P^2}{\sqrt{2 \log P}} \quad \text{and} \quad \widetilde{M_0} > \frac{0.0217 P^2}{\sqrt{2 \log P}}.$$

Then Lemmata 2.4 and 2.5 imply respectively that for $\varepsilon, \eta \in \{0,1\}$, one has

$$\int_0^1 |F_\eta(\alpha)^2 S_\varepsilon(\alpha)^4| d\alpha$$
$$< M_\eta^2 (\log P)^{9/2} \left(\frac{60 P^2}{M_\eta (\log P)^{1/2}} + \frac{P^4}{M_\eta^2} \left(\frac{500}{(\log P)^{3/2}} + \frac{15360}{(\log P)^{5/2}} \right) \right)$$
$$< M_\eta^2 (\log P)^{9/2} \left(\frac{60\sqrt{2}}{0.0508} + \frac{1000}{0.0508^2 (\log P)^{1/2}} + \frac{30720}{0.0508^2 (\log P)^{3/2}} \right)$$
$$< 45578 M_\eta^2 (\log P)^{9/2}, \tag{2.14}$$

and also, for $\zeta \in \{0,1\}$,

$$\int_0^1 |D_\zeta(\alpha)^2 S_1(\alpha)^2| d\alpha < \widetilde{M}_0^2 (\log P)^{9/2} \left(\frac{10 P^2}{\widetilde{M}_0 (\log P)^{1/2}} + \frac{45 P^4}{\widetilde{M}_0^2 (\log P)^{3/2}} \right)$$
$$< \widetilde{M}_0^2 (\log P)^{9/2} \left(\frac{10\sqrt{2}}{0.0217} + \frac{90}{0.0217^2 (\log P)^{1/2}} \right)$$
$$< 17953 \widetilde{M}_0^2 (\log P)^{9/2}. \tag{2.15}$$

Next write $S(\alpha) = \max\{|S_0(\alpha)|, |S_1(\alpha)|\}$. Then by Schwarz's inequality we obtain

$$|R(N; \mathfrak{m})| \leqslant \int_{\mathfrak{m}} |F_\eta(\alpha) D_\zeta(\alpha) S_0(\alpha)^{7-t} S_1(\alpha)^t| d\alpha$$
$$\leqslant \left(\sup_{\alpha \in \mathfrak{m}} S(\alpha) \right)^4 \left(\int_0^1 |F_\eta(\alpha)^2 S_\varepsilon(\alpha)^4| d\alpha \right)^{1/2}$$
$$\times \left(\int_0^1 |D_\zeta(\alpha)^2 S_1(\alpha)^2| d\alpha \right)^{1/2},$$

where ε is 0 or 1 according to whether $1 \leqslant t \leqslant 2$ or $3 \leqslant t \leqslant 7$. Then by applying Lemma 2.3 in combination with (2.14) and (2.15), we deduce that

$$|R(N; \mathfrak{m})| < 16.7^4 \sqrt{45578 \times 17953} M_\eta \widetilde{M}_0 P^{3.536} (\log P)^{5.5}.$$

In view of (2.12), it thus follows from Lemma 2.1 that for a certain positive number ν with $\nu < 64$, one has

$$R(N) \geqslant R(N; \mathfrak{M}) - |R(N; \mathfrak{m})| > 0.00021 M_\eta \widetilde{M}_0 P^4 (1 - E), \tag{2.16}$$

where
$$E = 1.0595 \times 10^{13} P^{-0.464} (\log P)^{5.5}. \tag{2.17}$$

A modest calculation reveals that the expression on the right hand side of (2.17) is less than 1 whenever $P > 7 \times 10^{52}$, and thus we deduce from (2.16) that $R(N) > 0$ whenever $P \geqslant 10^{53}$. On recalling (2.10) and the hypothesis $0 < \nu < 64$, we find that the latter condition on P is satisfied whenever $N \geqslant N_0$, where

$$N_0 = 16\nu(10^{53})^4 < 1.1 \times 10^{215}.$$

Thus we may conclude that whenever $N \geqslant 10^{216}$ and $16 \nmid N$, then N is indeed the sum of 16 biquadrates. The proof of Theorem 2 therefore follows on establishing Lemmata 2.1–2.5, this task being our primary concern in §§ 3-11.

CHAPTER 3

THE CARDINALITY OF THE SETS $\mathcal{M}_\eta(X)$

Our goal in this section is the proof of Lemma 2.2, which provides the explicit control of the distribution of integers of the shape $x^2 + xy + y^2$ essential to the main body of our argument. We begin by establishing some preliminary lemmata that provide basic information concerning the distribution of prime numbers. In this context, we remark that sharper versions of our Lemmata 3.1, 3.2 and 3.3 follow swiftly from the work of McCurley [17] and Rosser and Schoenfeld [19]. However, in order to avoid dependence on the heavy computations inherent in the latter work, we seek here to provide self-contained proofs of conclusions sufficient for the applications at hand.

Let $\Lambda(n)$ denote the von Mangoldt function, defined to be $\log p$ when n is a prime power p^r, and zero otherwise. Also, define

$$\psi(x) = \sum_{n \leqslant x} \Lambda(n).$$

Our first lemma provides an estimate for $\psi(x)$ via an argument of Chebyshev.

LEMMA 3.1. — *For $x \geqslant 41$, one has*

$$0.9212x - 5\log x < \psi(x) < 1.1056x + 3(\log x)^2.$$

Proof. — Define the function $f(t)$ by

$$f(t) = [t] - [t/2] - [t/3] - [t/5] + [t/30].$$

Then one readily verifies that $f(t)$ is equal to 0 or 1 for every real number t, and further that $f(t) = 1$ for $1 \leqslant t < 6$. Write

$$\Psi(x) = \log([x]!) - \log([x/2]!) - \log([x/3]!) - \log([x/5]!) + \log([x/30]!). \tag{3.1}$$

Then, on recalling the well-known formula

$$\sum_{n \leqslant x} \Lambda(n) \left[\frac{x}{n}\right] = \sum_{n \leqslant x} \log n = \log([x]!) \tag{3.2}$$

(see, for example, Theorem 3.12 of Apostol [**1**]), we deduce that
$$\psi(x) - \psi(x/6) \leqslant \Psi(x) \leqslant \psi(x). \tag{3.3}$$
An application of Euler's summation formula (see, for example, Theorem 3.1 of [**1**]) reveals that
$$\sum_{n \leqslant x} \log n = \int_1^x \log t \, dt + \int_1^x \frac{t - [t]}{t} dt - (x - [x]) \log x.$$
But for $x \geqslant 1$ one has the trivial estimates
$$0 \leqslant \int_1^x (t - [t]) t^{-1} dt \leqslant \log x$$
and
$$0 \leqslant (x - [x]) \log x \leqslant \log x,$$
and hence we deduce that
$$\left|\log([x]!) - (x \log x - x + 1)\right| \leqslant \log x. \tag{3.4}$$
On writing
$$c = \log(2^{1/2} 3^{1/3} 5^{1/5} 30^{-1/30}), \tag{3.5}$$
and combining (3.1) and (3.4), we deduce that for $x \geqslant 30$ one has
$$|\Psi(x) - cx + 1| \leqslant \log x + \log(x/2) + \log(x/3) + \log(x/5) + \log(x/30)$$
$$= 5 \log x - 2 \log 30. \tag{3.6}$$

Collecting together (3.3) and (3.6), we obtain the lower bound
$$\psi(x) \geqslant \Psi(x) \geqslant cx - 5 \log x. \tag{3.7}$$
On the other hand, the inequalities (3.3) and (3.6) yield also
$$\psi(x) - \psi(x/6) \leqslant \Psi(x) \leqslant cx + 5 \log x - 1 - 2 \log 30,$$
from which one deduces that
$$\psi(x) \leqslant \sum_{\substack{h \geqslant 0 \\ x/6^h \geqslant 30}} \left(\psi(x/6^h) - \psi(x/6^{h+1})\right) + \psi(30)$$
$$< c \sum_{h=0}^{\infty} \frac{x}{6^h} + \left(1 + \frac{\log(x/30)}{\log 6}\right)(5 \log x - 1 - 2 \log 30) + \psi(30).$$
A modest computation reveals that $\psi(30) < 28.48$, and thus a minor calculation demonstrates that for $x \geqslant 41$ one has
$$\psi(x) < \frac{6}{5} cx + \frac{5}{\log 6} (\log x)^2 - \frac{\log x}{\log 6} (5 \log 5 + 1 + 2 \log 30) + 35.5$$
$$< \frac{6}{5} cx + 3 (\log x)^2. \tag{3.8}$$
The number c defined by (3.5) satisfies $0.9212 < c < 0.9213$, and thus the lemma follows immediately from (3.7) and (3.8). □

We obtain a bound for the sum of reciprocals of the primes up to X via a familiar partial summation argument.

LEMMA 3.2. — *When $X \geqslant 10^{25}$, one has*
$$\sum_{p \leqslant X} \frac{1}{p} < \log \log X + 0.281.$$

Proof. — We begin by estimating the sum
$$B(x) = \sum_{p \leqslant x} \frac{\log p}{p}. \tag{3.9}$$

Observe first that in view of the formulae (3.2) and (3.4), when $x \geqslant 10^{25}$ one has
$$\sum_{n \leqslant x} \frac{\Lambda(n)}{n} \geqslant \frac{1}{x} \sum_{n \leqslant x} \Lambda(n) \left[\frac{x}{n}\right] \geqslant \log x - 1 - \frac{\log x - 1}{x}$$
$$> \log x - 1 - 10^{-23}. \tag{3.10}$$

Similarly, from (3.2), (3.4), and the conclusion of Lemma 3.1, when $x \geqslant 10^{25}$ one has
$$\sum_{n \leqslant x} \frac{\Lambda(n)}{n} \leqslant \frac{1}{x}\left(\sum_{n \leqslant x} \Lambda(n)\left[\frac{x}{n}\right] + \sum_{n \leqslant x} \Lambda(n)\right)$$
$$\leqslant \log x - 1 + \frac{\log x + 1}{x} + 1.1056 + \frac{3(\log x)^2}{x}$$
$$< \log x + 0.1056 + 10^{-21}. \tag{3.11}$$

In order to remove the contribution to the latter sums arising from the higher prime powers, we note next that
$$\sum_{p \leqslant \sqrt{x}} \sum_{\substack{m \geqslant 2 \\ p^m \leqslant x}} \frac{\log p}{p^m} = \sum_{p \leqslant \sqrt{x}} \frac{\log p}{p(p-1)} - \sum_{p \leqslant \sqrt{x}} \frac{\log p}{p^{m_p}(p-1)}, \tag{3.12}$$

where $m_p = [\log x / \log p]$. On making use of the trivial bound $p^{m_p}(p-1) \geqslant p^{m_p+1}/2 > x/2$, it follows that for $x \geqslant 10^{25}$ one has
$$0 \leqslant \sum_{p \leqslant \sqrt{x}} \frac{\log p}{p^{m_p}(p-1)} \leqslant \frac{2 \log \sqrt{x}}{x} \sum_{p \leqslant \sqrt{x}} 1 \leqslant \frac{\log x}{\sqrt{x}} < 10^{-10}. \tag{3.13}$$

Moreover, when $x \geqslant 10^{25}$ one has also
$$0 \leqslant \sum_{p > \sqrt{x}} \frac{\log p}{p(p-1)} < 2 \sum_{n > \sqrt{x}} \frac{\log n}{n^2} < 2 \int_{\sqrt{x}-1}^{\infty} \frac{\log t}{t^2} dt$$
$$= 2 \frac{\log(\sqrt{x}-1)+1}{\sqrt{x}-1} < 10^{-10}. \tag{3.14}$$

We now write
$$A_0 = \sum_{p} \sum_{m \geqslant 2} \frac{\log p}{p^m} = \sum_{p} \frac{\log p}{p(p-1)},$$

and conclude from (3.12), (3.13) and (3.14) that

$$A_0 - 10^{-9} < \sum_{p \leqslant \sqrt{x}} \sum_{\substack{m \geqslant 2 \\ p^m \leqslant x}} \frac{\log p}{p^m} \leqslant A_0. \tag{3.15}$$

Finally, defining the function $r(x)$ by means of the relation

$$B(x) = \log x + r(x), \tag{3.16}$$

we may conclude from (3.9), (3.10), (3.11) and (3.15) that for $x \geqslant 10^{25}$ one has

$$-A_0 - 1.0001 < r(x) < -A_0 + 0.1057. \tag{3.17}$$

By applying a partial summation argument along the lines applied in the proof of Theorem 7 of Ingham [**14**, Chapter I], one deduces from (3.16) that

$$\sum_{p \leqslant X} \frac{1}{p} = \log \log X + B_0 + \frac{r(X)}{\log X} - \int_X^\infty \frac{r(x)}{x(\log x)^2} dx, \tag{3.18}$$

where

$$B_0 = \gamma_0 + \sum_p \left(\log\left(1 - \frac{1}{p}\right) + \frac{1}{p} \right), \tag{3.19}$$

and $\gamma_0 = 0.5772...$ denotes Euler's constant. We obtain an upper bound for the infinite sum in (3.19) by the following simple device. We note first that the Riemann zeta function $\zeta(s)$ satisfies

$$\zeta(2) = \prod_p (1 - p^{-2})^{-1} = \frac{\pi^2}{6}. \tag{3.20}$$

Then on noting that for $0 < t < 1$, one has $\log(1-t) + t < \log(\sqrt{1-t^2})$, one finds that

$$\sum_p \left(\log\left(1 - \frac{1}{p}\right) + \frac{1}{p} \right)$$

$$< \sum_{p \leqslant 23} \left(\log\left(1 - \frac{1}{p}\right) + \frac{1}{p} - \frac{1}{2}\log\left(1 - \frac{1}{p^2}\right) \right) - \frac{1}{2}\log \zeta(2)$$

$$< -0.0668 - 0.2488 = -0.3156.$$

We therefore deduce from (3.19) that

$$B_0 < 0.2617. \tag{3.21}$$

Finally, on recalling (3.17) one finds that for $X \geqslant 10^{25}$ one has

$$\frac{r(X)}{\log X} - \int_X^\infty \frac{r(x)}{x(\log x)^2} dx < \frac{-A_0 + 0.1057}{\log X} + \int_X^\infty \frac{A_0 + 1.0001}{x(\log x)^2} dx$$

$$= \frac{1.1058}{\log X} < 0.0193.$$

The conclusion of the lemma now follows by substituting (3.21) and the latter estimate into (3.18). □

We next consider the distribution of primes in arithmetic progressions modulo 6. We define
$$\pi_1(x) = \sum_{\substack{p \leqslant x \\ p \equiv 1 \pmod{6}}} 1.$$

LEMMA 3.3. — *When $x \geqslant 10^{20}$, one has*
$$\pi_1(x) \geqslant 0.3687 \frac{x}{\log x}.$$

Proof. — We bound $\pi_1(x)$ from below by employing a lower bound for $\psi_1(x)$, which we define by
$$\psi_1(x) = \sum_{\substack{1 \leqslant n \leqslant x \\ n \equiv 1 \pmod{6}}} \Lambda(n).$$

Denote by χ_1 the non-trivial character modulo 6, and write χ_0 for the corresponding trivial character. Define also $\psi(x, \chi)$ for $\chi = \chi_0, \chi_1$ by
$$\psi(x, \chi) = \sum_{1 \leqslant n \leqslant x} \chi(n) \Lambda(n).$$

Then by using simple properties of characters, one finds that
$$2\psi_1(x) = \psi(x, \chi_0) + \psi(x, \chi_1). \tag{3.22}$$

We observe that
$$\psi(x, \chi_0) \leqslant \psi(x) \leqslant \psi(x, \chi_0) + \sum_{1 < 2^r \leqslant x} \log 2 + \sum_{1 < 3^r \leqslant x} \log 3$$
$$\leqslant \psi(x, \chi_0) + 2 \log x,$$

whence
$$\psi(x) - 2\log x \leqslant \psi(x, \chi_0) \leqslant \psi(x). \tag{3.23}$$

It is useful at this point to derive some simple properties of alternating series associated with certain character sums. We assume first that $f(t)$ is a function that is monotone decreasing for $t > 0$, and satisfies the condition that $f(t) = 0$ when t exceeds some real number x with $x \geqslant 1$. Then on noting that $-f(6m-1) + f(6m+1) \leqslant 0$ and $f(6m+1) - f(6m+5) \geqslant 0$ for every integer m, one finds that the alternating series
$$\sum_{1 \leqslant n \leqslant x} \chi_1(n) f(n) = f(1) - f(5) + f(7) - f(11) + f(13) - f(17) + \cdots$$

satisfies the property that for each natural number n_0, one has
$$f(1) \geqslant \sum_{1 \leqslant n \leqslant x} \chi_1(n) f(n) \geqslant \sum_{n=1}^{6n_0 - 1} \chi_1(n) f(n). \tag{3.24}$$

Suppose next that $f(t)$ is a function that is monotone and non-negative in the range $x \leqslant t \leqslant y$. Then one obtains in a similar manner the upper bound

$$\left| \sum_{x < n \leqslant y} \chi_1(n) f(n) \right| \leqslant \max\{f(x), f(y)\}. \tag{3.25}$$

Returning now to the main task of bounding $\psi_1(x)$ from below, we begin by noting that

$$\sum_{1 \leqslant n \leqslant x} \chi_1(n) \log n = \sum_{1 \leqslant n \leqslant x} \chi_1(n) \sum_{m|n} \Lambda(m)$$

$$= \sum_{1 \leqslant d \leqslant x} \chi_1(d) \sum_{1 \leqslant m \leqslant x/d} \chi_1(m) \Lambda(m),$$

so that an application of (3.25) with $f(t) = \log t$ yields the bound

$$\left| \sum_{1 \leqslant d \leqslant x} \chi_1(d) \psi(x/d, \chi_1) \right| = \left| \sum_{1 \leqslant n \leqslant x} \chi_1(n) \log n \right| \leqslant \log x. \tag{3.26}$$

Next applying the first inequality in (3.24) with $f(t) = \psi_1(x/t)$, and then recalling (3.22), we find that

$$\psi_1(x) \geqslant \sum_{1 \leqslant n \leqslant x} \chi_1(n) \psi_1(x/n)$$

$$= \frac{1}{2} \left(\sum_{1 \leqslant n \leqslant x} \chi_1(n) \psi(x/n, \chi_0) + \sum_{1 \leqslant n \leqslant x} \chi_1(n) \psi(x/n, \chi_1) \right).$$

Estimating the second sum in the latter inequality by means of (3.26), and the first by applying (3.24) with $f(t) = \psi(x/t, \chi_0)$ and $n_0 = 4$, we deduce that

$$\psi_1(x) \geqslant \frac{1}{2} \sum_{n=1}^{23} \chi_1(n) \psi(x/n, \chi_0) - \frac{1}{2} \log x.$$

Consequently, it follows from (3.23) and Lemma 3.1 that when $x \geqslant 10^{20}$, one has

$$\psi_1(x) \geqslant \frac{1}{2} \left(\psi(x) - \psi(x/5) + \psi(x/7) - \psi(x/11) \right.$$
$$\left. + \psi(x/13) - \psi(x/17) + \psi(x/19) - \psi(x/23) \right) - \frac{9}{2} \log x$$
$$> \frac{x}{2} \left(0.9212 \left(1 + \frac{1}{7} + \frac{1}{13} + \frac{1}{19} \right) - 1.1056 \left(\frac{1}{5} + \frac{1}{11} + \frac{1}{17} + \frac{1}{23} \right) \right)$$
$$- 6(\log x)^2 - 15(\log x)$$
$$> 0.368705 x. \tag{3.27}$$

In order to establish the desired lower bound for $\pi_1(x)$, we note first that
$$\sum_{m \geqslant 2} \sum_{\substack{p^m \leqslant x \\ p^m \equiv 1 \pmod 6}} \log p \leqslant \log x \sum_{p \leqslant \sqrt{x}} 1 \leqslant \sqrt{x} \log x.$$
It therefore follows from (3.27) that for $x \geqslant 10^{20}$, one has
$$\pi_1(x) \geqslant \frac{1}{\log x}\left(\psi_1(x) - \sum_{m \geqslant 2} \sum_{\substack{p^m \leqslant x \\ p^m \equiv 1 \pmod 6}} \log p\right)$$
$$> (0.368705 x - \sqrt{x} \log x)/\log x > 0.3687 x/\log x.$$

This completes the proof of the lemma. □

LEMMA 3.4. — *When $X \geqslant 10^{25}$, one has*
$$\sum_{\substack{p \leqslant X \\ p \equiv -1 \pmod 6}} \frac{1}{p} < \frac{1}{2} \log \log X - 0.195.$$

Proof. — Before proceeding to the main part of our argument, we require several preliminary estimates. Let χ_0 and χ_1 denote the trivial and non-trivial characters modulo 6, as in the proof of the previous lemma, and denote by $L(s, \chi)$ the Dirichlet L-function
$$L(s, \chi) = \sum_{n=1}^{\infty} \chi(n) n^{-s}$$
associated with the character χ. For the sake of concision, it is convenient to write
$$L = L(1, \chi_1) = \sum_{n=1}^{\infty} \frac{\chi_1(n)}{n}, \quad \text{and} \quad L' = L'(1, \chi_1) = -\sum_{n=1}^{\infty} \frac{\chi_1(n) \log n}{n}.$$
Let χ^* denote the non-trivial character modulo 3, so that χ^* is the primitive character which induces χ_1. Then it is known that $L(1, \chi^*) = \pi/(3\sqrt{3})$ (see, for example, Theorems 12.11 and 12.20 of Apostol [1]), from which it follows that
$$L = \prod_p (1 - \chi_1(p)/p)^{-1} = \left(1 + \frac{1}{2}\right) L(1, \chi^*) = \frac{\pi}{2\sqrt{3}}. \tag{3.28}$$

So far as L' is concerned, we remark that it is possible to confirm that
$$L' = \frac{\pi}{2\sqrt{3}}\left(3 \log \Gamma(2/3) - 3 \log \Gamma(1/3) + (2/3) \log 2 + \log \pi + \gamma_0\right),$$
where we use Γ to denote the gamma function, and γ_0 again denotes Euler's constant. Instead of establishing the latter formula, we note that
$$L' = \sum_{k=1}^{\infty} \left(\frac{\log(6k-1)}{6k-1} - \frac{\log(6k+1)}{6k+1}\right).$$

Plainly, one has
$$\sum_{k=17}^{\infty}\left(\frac{\log(6k-1)}{6k-1}-\frac{\log(6k+1)}{6k+1}\right) < \int_{16}^{\infty}\left(\frac{\log(6t-1)}{6t-1}-\frac{\log(6t+1)}{6t+1}\right)dt$$
$$=\frac{(\log 97)^2-(\log 95)^2}{12} < 0.016,$$
and a straightforward calculation confirms that
$$\sum_{k=1}^{16}\left(\frac{\log(6k-1)}{6k-1}-\frac{\log(6k+1)}{6k+1}\right) < 0.109.$$
Thus one obtains the estimate
$$0 < L' < 1/8. \tag{3.29}$$

We next estimate some partial sums required in our subsequent deliberations. First, we deduce from the argument leading to (3.24) that for each $x \geqslant 1$, one has
$$1 \geqslant \sum_{1 \leqslant n \leqslant x} \frac{\chi_1(n)}{n} \geqslant 1 - \frac{1}{5} = \frac{4}{5}.$$
Thus, in view of (3.28) and (3.25), one deduces that
$$\left|L - \sum_{1 \leqslant m \leqslant x} \frac{\chi_1(m)}{m}\right| = \left|\sum_{m>x} \frac{\chi_1(m)}{m}\right| \leqslant \min\left\{\frac{1}{x}, \frac{\pi}{2\sqrt{3}} - \frac{4}{5}\right\}. \tag{3.30}$$
We also find from (3.25) that
$$\left|L' + \sum_{1 \leqslant m \leqslant x} \frac{\chi_1(m)\log m}{m}\right| = \left|\sum_{m>x} \frac{\chi_1(m)\log m}{m}\right| \leqslant \frac{\log x}{x}. \tag{3.31}$$
Write $\mu(n)$ for the Möbius function, and observe next that
$$\sum_{1 \leqslant md \leqslant x} \frac{\chi_1(md)\mu(d)}{md} = \sum_{1 \leqslant n \leqslant x} \frac{\chi_1(n)}{n} \sum_{d|n} \mu(d) = 1.$$
Consequently, on making use of (3.30), we deduce that for $x \geqslant 10^{10}$, one has
$$\left|L\sum_{1 \leqslant d \leqslant x} \frac{\chi_1(d)\mu(d)}{d} - 1\right| = \left|\sum_{1 \leqslant d \leqslant x} \frac{\chi_1(d)\mu(d)}{d}\left(L - \sum_{1 \leqslant m \leqslant x/d} \frac{\chi_1(m)}{m}\right)\right|$$
$$\leqslant \sum_{1 \leqslant d \leqslant x/9} \frac{1}{d}\left(\frac{x}{d}\right)^{-1} + \left(\frac{\pi}{2\sqrt{3}} - \frac{4}{5}\right) \sum_{x/9 < d \leqslant x} \frac{1}{d}$$
$$\leqslant \frac{1}{9} + 0.1069 \int_{x/9-1}^{x} \frac{dt}{t}$$
$$= \frac{1}{9} + 0.1069 \log\left(\frac{9x}{x-9}\right) < 0.346. \tag{3.32}$$

We next recall that
$$\Lambda(n) = \sum_{md=n} \mu(d) \log m,$$
and hence obtain the relation
$$\sum_{1 \leqslant n \leqslant x} \frac{\chi_1(n)\Lambda(n)}{n} = \sum_{1 \leqslant d \leqslant x/5} \frac{\chi_1(d)\mu(d)}{d} \sum_{1 \leqslant m \leqslant x/d} \frac{\chi_1(m) \log m}{m},$$

Here we have observed that the innermost sum in the last expression is zero for $d > x/5$. It therefore follows that

$$\sum_{1 \leqslant n \leqslant x} \frac{\chi_1(n)\Lambda(n)}{n} + \frac{L'}{L} = \sum_{1 \leqslant d \leqslant x/5} \frac{\chi_1(d)\mu(d)}{d} \left(\sum_{1 \leqslant m \leqslant x/d} \frac{\chi_1(m) \log m}{m} + L' \right)$$
$$- \frac{L'}{L} \left(L \sum_{1 \leqslant d \leqslant x/5} \frac{\chi_1(d)\mu(d)}{d} - 1 \right). \quad (3.33)$$

But by (3.31) and (3.4), when $x \geqslant 10^{10}$ one has

$$\sum_{1 \leqslant d \leqslant x/5} \frac{1}{d} \left| \sum_{1 \leqslant m \leqslant x/d} \frac{\chi_1(m) \log m}{m} + L' \right| \leqslant \sum_{1 \leqslant d \leqslant x/5} \frac{\log(x/d)}{x}$$
$$\leqslant \frac{1}{x} \left(\frac{x}{5} \log x - \left(\frac{x}{5} \log\left(\frac{x}{5}\right) - \frac{x}{5} + 1 - \log\left(\frac{x}{5}\right) \right) \right)$$
$$= \frac{1 + \log 5}{5} + \frac{\log(x/5) - 1}{x} < 0.5219. \quad (3.34)$$

Finally, on recalling (3.28) and (3.29), and substituting (3.32) and (3.34) into (3.33) we conclude that for $x \geqslant 5 \times 10^{10}$, one has

$$\left| \sum_{1 \leqslant n \leqslant x} \frac{\chi_1(n)\Lambda(n)}{n} + \frac{L'}{L} \right| < 0.5219 + \frac{1}{8} \cdot \frac{2\sqrt{3}}{\pi} 0.346 < 0.5696. \quad (3.35)$$

It is now time to estimate the contribution of the higher powers of primes to the sum central to this lemma. To this end, we put

$$A_1 = \sum_p \sum_{m \geqslant 2} \frac{\chi_1(p^m) \log p}{p^m} = \sum_p \frac{\log p}{p(p - \chi_1(p))},$$

and note that

$$\sum_{\substack{p \leqslant \sqrt{x} \\ p^m \leqslant x}} \sum_{m \geqslant 2} \frac{\chi_1(p^m) \log p}{p^m} = A_1 - \sum_{p > \sqrt{x}} \frac{\log p}{p(p - \chi_1(p))} - \sum_{p \leqslant \sqrt{x}} \frac{\chi_1(p)^{m_p+1} \log p}{p^{m_p}(p - \chi_1(p))},$$

where $m_p = [\log x / \log p]$. On recalling (3.13) and (3.14), we therefore obtain for $x \geqslant 10^{25}$ the bound

$$\left| \sum_{1 \leqslant n \leqslant x} \frac{\chi_1(n) \Lambda(n)}{n} - \sum_{p \leqslant x} \frac{\chi_1(p) \log p}{p} - A_1 \right|$$

$$= \left| \sum_{\substack{p \leqslant \sqrt{x} \\ p^m \leqslant x}} \sum_{m \geqslant 2} \frac{\chi_1(p^m) \log p}{p^m} - A_1 \right|$$

$$\leqslant \sum_{p > \sqrt{x}} \frac{\log p}{p(p-1)} + \sum_{p \leqslant \sqrt{x}} \frac{\log p}{p^{m_p}(p-1)} < 10^{-9}. \qquad (3.36)$$

Write $c_j = -L'/L - A_1 + (-1)^j \cdot 0.57$ for $j = 1, 2$. Then it follows from (3.35) and (3.36) that for $x \geqslant 10^{25}$, one has

$$c_1 < \sum_{p \leqslant x} \frac{\chi_1(p) \log p}{p} < c_2,$$

Consequently, whenever $y > x \geqslant 10^{25}$, it follows via partial summation that

$$\sum_{x < p \leqslant y} \frac{\chi_1(p)}{p} = \frac{1}{\log y} \sum_{p \leqslant y} \frac{\chi_1(p) \log p}{p} - \frac{1}{\log x} \sum_{p \leqslant x} \frac{\chi_1(p) \log p}{p}$$

$$+ \int_x^y \sum_{p \leqslant t} \frac{\chi_1(p) \log p}{p} \cdot \frac{dt}{t(\log t)^2}$$

$$< \frac{c_2}{\log y} - \frac{c_1}{\log x} + c_2 \int_x^y \frac{dt}{t(\log t)^2} = \frac{c_2 - c_1}{\log x}.$$

We therefore conclude that when $x \geqslant 10^{25}$, one has

$$\sum_{p > x} \frac{\chi_1(p)}{p} \leqslant \frac{1.14}{\log x} < 0.02. \qquad (3.37)$$

Next we examine the sum

$$A_2 = \sum_p \frac{\chi_1(p)}{p}.$$

Note first that the formulae (3.20) and (3.28) yield

$$\sum_{p \geqslant 5} \log(1 - \chi_1(p)/p) - \frac{1}{2} \sum_{p \geqslant 5} \log(1 - 1/p^2)$$

$$= \log(1/L) + \frac{1}{2} \log\big((1 - 2^{-2})(1 - 3^{-2})\zeta(2)\big)$$

$$= \log(2/\sqrt{3}).$$

Thus, on writing

$$F(p) = \frac{\chi_1(p)}{p} + \log\Big(1 - \frac{\chi_1(p)}{p}\Big) - \frac{1}{2} \log\Big(1 - \frac{1}{p^2}\Big),$$

MÉMOIRES DE LA SMF 100

we arrive at the formula

$$A_2 = \sum_{p \geqslant 5} F(p) - \log(2/\sqrt{3}). \tag{3.38}$$

We estimate the sum on the right hand side of (3.38) by appealing to the Taylor expansion

$$\log(1-t) = -\sum_{m=1}^{\infty} \frac{t^m}{m},$$

which is valid whenever $|t| < 1$. Thus, for each prime p with $p \geqslant 5$, we find that

$$F(p) = -\sum_{m=2}^{\infty} \frac{\chi_1(p)^m}{mp^m} + \frac{1}{2}\sum_{n=1}^{\infty} \frac{1}{np^{2n}} = -\chi_1(p) \sum_{\substack{m \geqslant 3 \\ m \equiv 1 \,(\text{mod } 2)}} \frac{1}{mp^m}.$$

In particular, when $p \equiv 5 \pmod{6}$ it is apparent that $F(p) > 0$, and when $p \equiv 1 \pmod{6}$ one has

$$F(p) \geqslant -\frac{1}{3}\sum_{l=1}^{\infty} p^{-1-2l} = -\frac{1}{3p(p^2-1)} = \frac{1}{6}\Big(\frac{1}{p(p+1)} - \frac{1}{p(p-1)}\Big).$$

Consequently,

$$\sum_{p>43} F(p) > \frac{1}{6} \sum_{\substack{p>43 \\ p \equiv 1 \,(\text{mod } 6)}} \Big(\frac{1}{p(p+1)} - \frac{1}{p(p-1)}\Big)$$

$$> \frac{1}{6} \sum_{n \geqslant 61} \Big(\frac{1}{n(n+1)} - \frac{1}{n(n-1)}\Big) = -\frac{1}{6 \cdot 61 \cdot 60}$$

$$> -0.0000456, \tag{3.39}$$

while a direct calculation yields

$$\sum_{5 \leqslant p \leqslant 43} F(p) > 0.00189. \tag{3.40}$$

We therefore conclude from (3.38), (3.39) and (3.40) that

$$A_2 > \log(\sqrt{3}/2) + 0.00189 - 0.0000456 > -0.142,$$

whence by (3.37), whenever $x \geqslant 10^{25}$, one has

$$\sum_{p \leqslant x} \frac{\chi_1(p)}{p} = A_2 - \sum_{p>x} \frac{\chi_1(p)}{p} > -0.162. \tag{3.41}$$

We are at last positioned to deliver the conclusion of the lemma. On combining the conclusion of Lemma 3.2 with (3.41), we find that when $x \geqslant 10^{25}$, one has

$$\sum_{\substack{p \leqslant x \\ p \equiv -1 \pmod{6}}} \frac{1}{p} = \frac{1}{2}\left(\sum_{5 \leqslant p \leqslant x} \frac{1}{p} - \sum_{p \leqslant x} \frac{\chi_1(p)}{p}\right)$$

$$< \frac{1}{2}\left(\log\log x + 0.281 - \frac{1}{2} - \frac{1}{3}\right) + \frac{1}{2}(0.162)$$

$$< \frac{1}{2}\log\log x - 0.195,$$

and this completes the proof of the lemma. □

At last we are equipped to dispose of the proof of Lemma 2.2.

Proof of Lemma 2.2. — We begin by making the crucial observation, familiar from the theory of binary quadratic forms, that a natural number n is represented by $x^2 + xy + y^2$, with integers x and y, if and only if n satisfies the condition that whenever $p|n$ with $p \equiv 2 \pmod{3}$, then for some natural number h one has $p^{2h} \| n$. Denote by $\mathcal{N}(X)$ the set of odd integers up to X that are represented by $x^2 + xy + y^2$, so that $\mathcal{N}(X)$ is the union of $\mathcal{M}_1(X)$ and the set of odd squares up to X. Then in view of the above comments, every natural number n with $n \leqslant X$ can be uniquely expressed in the form $n = 2^k l m$, where $k \geqslant 0$, $l \in \mathcal{N}(X)$, and m is a product of distinct prime numbers congruent to -1 modulo 6. We seek to obtain a lower bound for a weighted sum over the elements of $\mathcal{N}(X)$ by observing that our last remark ensures that

$$\left(\sum_{k=0}^{\infty} \frac{1}{2^k}\right)\left(\prod_{\substack{p \leqslant X \\ p \equiv -1 \pmod{6}}}\left(1 + \frac{1}{p}\right)\right) \sum_{m \in \mathcal{N}(X)} \frac{1}{m} > \sum_{n \leqslant X} \frac{1}{n}$$

$$> \int_1^X \frac{dt}{t} = \log X. \quad (3.42)$$

But when $0 < t < 1$, one has $\log(1+t) < t$, and thus Lemma 3.4 shows that whenever $X \geqslant 10^{25}$,

$$\log\left(\prod_{\substack{p \leqslant X \\ p \equiv -1 \pmod{6}}}\left(1 + \frac{1}{p}\right)\right) < \sum_{\substack{p \leqslant X \\ p \equiv -1 \pmod{6}}} \frac{1}{p} < \frac{1}{2}\log\log X - 0.195.$$

Consequently, it follows from (3.42) that for $X \geqslant 10^{25}$ one has

$$\sum_{m \in \mathcal{N}(X)} \frac{1}{m} > \frac{1}{2}e^{0.195}\sqrt{\log X} > 0.6076\sqrt{\log X}. \quad (3.43)$$

We suppose next that $X \geqslant 10^{56}$, and write $Y = X^{13/28}$. Plainly, one has $Y \geqslant 10^{26}$. In view of our earlier remarks concerning the set of integers represented by the

quadratic form $x^2 + xy + y^2$, it is evident that
$$\mathcal{M}_1(X) \supset \{mp : m \in \mathcal{N}(Y), Y < p \leqslant X/m, p \equiv 1 \pmod 6\},$$
whence it follows that
$$\operatorname{card}(\mathcal{M}_1(X)) \geqslant \sum_{m \in \mathcal{N}(Y)} (\pi_1(X/m) - \pi_1(Y)).$$
Note here that $X/m \geqslant X^{15/28} > Y > 10^{20}$. Then on applying Lemma 3.3 in combination with the trivial upper bound $\pi_1(Y) \leqslant 1 + Y/6$, we obtain
$$\operatorname{card}(\mathcal{M}_1(X)) > \sum_{m \in \mathcal{N}(Y)} \left(\frac{0.3687 X}{m \log X} - \frac{Y}{6} - 1 \right)$$
$$\geqslant \frac{0.3687 X}{\log X} \sum_{m \in \mathcal{N}(Y)} \frac{1}{m} - \frac{Y^2}{6} - Y,$$
whence, on recalling (3.43), we find that $\operatorname{card}(\mathcal{M}_1(X))$ is greater than
$$\left(0.3687 \left(0.6076 \sqrt{\frac{13}{28}} \right) - \sqrt{\log X} \left(\frac{1}{6} X^{-1/14} + X^{-15/28} \right) \right) \frac{X}{\sqrt{\log X}}.$$
We therefore conclude that when $X \geqslant 10^{56}$, one has
$$\operatorname{card}(\mathcal{M}_1(X)) > 0.15245 X / \sqrt{\log X}. \tag{3.44}$$

In order to obtain a satisfactory lower bound for $\operatorname{card}(\mathcal{M}_0(X))$, we have only to remark that the sets
$$\{4^k m : m \in \mathcal{M}_1(4^{-k} X)\}$$
are pairwise disjoint for $k \geqslant 1$, and moreover each such set is plainly contained in $\mathcal{M}_0(X)$. Furthermore, since $4^{-6} X > 10^{56}$ whenever $X \geqslant 10^{60}$, we deduce from (3.44) that whenever the latter condition holds, one has
$$\operatorname{card}(\mathcal{M}_0(X)) \geqslant \sum_{k=1}^{6} \operatorname{card}(\mathcal{M}_1(4^{-k} X))$$
$$> \frac{0.15245 X}{\sqrt{\log X}} \sum_{k=1}^{6} 4^{-k} > 0.0508 \frac{X}{\sqrt{\log X}}. \tag{3.45}$$

The conclusion of Lemma 2.2 is immediate from (3.44) and (3.45). □

CHAPTER 4

AN AUXILIARY SINGULAR INTEGRAL

We avoid serious computations in our estimation of the singular integral by exploiting a probabilistic interpretation to obtain a simple lower bound, along the lines applied by Deshouillers in §2.1 of [6]. We first prepare an auxiliary lemma, and for this we require some notation. Write

$$J(\beta) = \int_1^2 e(\beta z^4) dz, \tag{4.1}$$

and, when m is a natural number with $m \geqslant 2$, define

$$K_m(\xi) = \int_{-\infty}^{\infty} J(\beta)^m e(-\xi\beta) d\beta. \tag{4.2}$$

It is simple to show that $K_m(\xi)$ is absolutely convergent for $m \geqslant 2$ (see, for example, inequality (4.3) below). We note for future reference that in order to establish the lower bound for the major arc contribution recorded in Lemma 2.1, it suffices to obtain a numerical lower bound for $K_7(\xi)$ holding uniformly for ξ in a suitable interval.

LEMMA 4.1. — *Suppose that δ is a positive number, and that ξ and ξ' are real numbers with $|\xi - \xi'| \leqslant \delta$. Then whenever $m \geqslant 3$, one has*

$$|K_m(\xi) - K_m(\xi')| \leqslant \frac{m\delta}{8(m-2)\pi}.$$

Proof. — When β is non-zero, it follows by partial integration that

$$J(\beta) = \frac{e(16\beta)}{64\pi i \beta} - \frac{e(\beta)}{8\pi i \beta} + \frac{3}{8\pi i \beta} \int_1^2 z^{-4} e(\beta z^4) dz,$$

and thus we obtain the estimate

$$|J(\beta)| \leqslant \frac{1}{64\pi|\beta|} + \frac{1}{8\pi|\beta|} + \frac{3}{8\pi|\beta|} \int_1^2 z^{-4} dz = \frac{1}{4\pi|\beta|}.$$

Combining the latter estimate with the trivial bound $|J(\beta)| \leqslant 1$, we deduce that

$$|J(\beta)| \leqslant \min\{1, (4\pi|\beta|)^{-1}\}. \tag{4.3}$$

On the other hand, it is apparent that the hypotheses of the lemma imply that
$$|e(-\xi\beta) - e(-\xi'\beta)| = |1 - e((\xi - \xi')\beta)|$$
$$= 2|\sin(\pi(\xi - \xi')\beta)| \leqslant 2\pi\delta|\beta|.$$

Thus, on making use of (4.3), we obtain the upper bounds
$$\left|\int_{|\beta| \leqslant (4\pi)^{-1}} J(\beta)^m (e(-\xi\beta) - e(-\xi'\beta)) d\beta\right| \leqslant 4\pi\delta \int_0^{(4\pi)^{-1}} \beta \, d\beta = \frac{\delta}{8\pi},$$

and, when $m \geqslant 3$,
$$\left|\int_{|\beta| \geqslant (4\pi)^{-1}} J(\beta)^m (e(-\xi\beta) - e(-\xi'\beta)) d\beta\right| \leqslant 4\pi\delta \int_{(4\pi)^{-1}}^{\infty} \beta (4\pi\beta)^{-m} d\beta$$
$$= \frac{\delta}{4(m-2)\pi}.$$

The conclusion of the lemma is immediate on combining the last two inequalities. $\square$

We now make use of the promised probabilistic interpretation.

LEMMA 4.2. — *Suppose that $m \geqslant 3$. Then there exists a real number $\nu = \nu(m)$ satisfying the inequalities*
$$\frac{31m}{5} - \frac{\sqrt{12378m}}{15} + \frac{1}{8} \leqslant \nu \leqslant \frac{31m}{5} + \frac{\sqrt{12378m}}{15} + \frac{1}{8},$$
with the property that whenever $\nu - \frac{1}{4} \leqslant \xi \leqslant \nu$, one has
$$K_m(\xi) \geqslant \frac{5}{\sqrt{12378m}} - \frac{m}{64(m-2)\pi}.$$

Proof. — Following the method of Deshouillers [6, §2.1], we consider m independent random variables $X_1, \ldots, X_m$ that are uniformly distributed on the interval $[1,2]$. On considering Fourier transforms and their inverses, as explained in [6] (see also §2.1 of [9]), we find that the integral $K_m(\xi)$ coincides with the density of the random variable $Z_m = X_1^4 + \cdots + X_m^4$ at ξ. Denote by $\mu = \mu(m)$ and $\sigma = \sigma(m)$, respectively, the mean and standard deviation of Z_m. Then we find that
$$\mu = m \int_1^2 X^4 dX = \frac{31m}{5}, \tag{4.4}$$

and
$$\sigma^2 = m \left(\int_1^2 X^8 dX - \left(\frac{31}{5}\right)^2 \right) = \frac{4126m}{225}. \tag{4.5}$$

By the Bienaymé-Chebyshev theorem, therefore, one has
$$\int_{\mu - \sqrt{3}\sigma}^{\mu + \sqrt{3}\sigma} K_m(\xi) d\xi \geqslant 1 - \left(\frac{1}{\sqrt{3}}\right)^2 = \frac{2}{3},$$

whence it follows that there exists a real number ξ_m, with
$$\mu - \sqrt{3}\sigma \leqslant \xi_m \leqslant \mu + \sqrt{3}\sigma, \tag{4.6}$$
such that
$$K_m(\xi_m) \geqslant \frac{1}{2\sqrt{3}\sigma} \cdot \frac{2}{3} = \frac{5}{\sqrt{12378m}}.$$
We may thus conclude from Lemma 4.1 that whenever $|\xi - \xi_m| \leqslant 1/8$, one has
$$K_m(\xi) \geqslant K_m(\xi_m) - |K_m(\xi_m) - K_m(\xi)|$$
$$\geqslant \frac{5}{\sqrt{12378m}} - \frac{m}{64(m-2)\pi}.$$
On setting $\nu = \nu(m) = \xi_m + 1/8$, the conclusion of the lemma is now immediate from (4.4)-(4.6). □

We close this section by extracting from Lemma 4.2 the conclusion relevant to our discussion in §7 involved in the proof of Lemma 2.1.

COROLLARY 4.3. — *There exists a real number ν, with $23 < \nu < 64$, satisfying the property that whenever ξ is a real number with $\nu - 1/4 \leqslant \xi \leqslant \nu$, then one has $K_7(\xi) \geqslant 0.01$.*

Proof. — The conclusion of the corollary is immediate from Lemma 4.2, on setting $m = 7$. □

We remark that numerical integration can be applied, with some computational expense, to establish that $K_7(\xi) > 0.0345$ for $41.5 \leqslant \xi \leqslant 42.5$.

CHAPTER 5

ESTIMATES FOR COMPLETE EXPONENTIAL SUMS

In advance of our discussion of the singular series, we prepare some preliminary estimates associated with the complete exponential sums

$$S(q, a) = \sum_{h=1}^{q} e\left(\frac{a}{q}h^4\right) \tag{5.1}$$

and

$$G_\varepsilon(q, a) = \sum_{h=1}^{q} e\left(\frac{a}{q}(2h + \varepsilon)^4\right) \qquad (\varepsilon = 0,\ 1). \tag{5.2}$$

In this context, we note that for odd q, one has

$$G_0(q, a) = G_1(q, a) = S(q, a). \tag{5.3}$$

In order to describe the known estimates for the above exponential sums, we define the function $c(q)$ for prime powers q as follows. We write

$$c(2) = 0, \quad c(2^2) = \sqrt{2}, \quad c(2^3) = \sqrt{2 + \sqrt{2}},\ c(2^4) = \sqrt{2 + \sqrt{2 + \sqrt{2}}},$$
$$c(5) = 1.32,\ c(13) = 1.138,\ c(17) = 1.269, \qquad c(41) = 1.142,$$

and define $c(p) = 1$ for the remaining odd primes p. We next define the multiplicative function $\kappa(q)$ by taking

$$\kappa(2^{4u+v}) = \begin{cases} 1, & \text{when } u = 0 \text{ and } 1 \leqslant v \leqslant 4, \\ 2^{-u}c(2^v), & \text{when } u \geqslant 1 \text{ and } 1 \leqslant v \leqslant 4, \end{cases} \tag{5.4}$$

and, when p is an odd prime, by defining $\kappa(p^h)$ through the relations

$$\kappa(p) = \begin{cases} p^{-1/2}, & \text{when } p \equiv 3 \pmod{4}, \\ \min\{3p^{-1/2},\ c(p)p^{-1/4}\}, & \text{when } p \equiv 1 \pmod{4}, \end{cases} \tag{5.5}$$

and

$$\kappa(p^{4u+v}) = \begin{cases} p^{-u}\kappa(p), & \text{when } u \geqslant 0 \text{ and } v = 1, \\ p^{-u-1}, & \text{when } u \geqslant 0 \text{ and } 2 \leqslant v \leqslant 4. \end{cases} \tag{5.6}$$

In view of the assumed multiplicative behaviour of $\kappa(q)$, these relations define $\kappa(q)$ for all natural numbers q.

We remark that $3p^{-1/2} < c(p)p^{-1/4}$ if, and only if, one has $p \geqslant 83$. Furthermore, for each natural number l one has $\kappa(2^l) \leqslant 2c(2^4)2^{-l/4}$, and, when p is an odd prime, one has also $\kappa(p^l) \leqslant c(p)p^{-l/4}$. We therefore deduce that for each natural number q, one has

$$\kappa(q) \leqslant 2c(2^4)c(5)c(13)c(17)c(41)q^{-1/4} < 9q^{-1/4}. \tag{5.7}$$

LEMMA 5.1. — *Let p be an odd prime, and suppose that l is a natural number. Then whenever $a \in \mathbb{Z}$ satisfies $(p,a) = 1$, one has*

$$|S(p^l, a)| \leqslant p^l \kappa(p^l).$$

Proof. — We begin by writing $l = 4u + v$ with $u \geqslant 0$ and $1 \leqslant v \leqslant 4$. Then according to Lemma 4.4 of Vaughan [21], one has

$$S(p^{4u+v}, a) = p^{3u} S(p^v, a), \tag{5.8}$$

and when $2 \leqslant v \leqslant 4$, one has $S(p^v, a) = p^{v-1}$. Thus we find that the conclusion of the lemma is immediate when $2 \leqslant v \leqslant 4$.

We next turn to the cases in which $v = 1$, noting initially that Lemma 4.3 of Vaughan [21] establishes the bound $|S(p,a)| \leqslant ((p-1,4)-1)\sqrt{p}$. The latter estimate suffices to establish the lemma whenever $p \equiv 3 \pmod{4}$, and also when $p \equiv 1 \pmod{4}$ and $p \geqslant 83$. In the remaining cases we extract the bound $|S(p,a)| \leqslant c(p)p^{3/4}$ from the argument of the proof of Lemma 2.2.2 of Deshouillers and Dress [9] (see especially the inequalities (2.2.7) and (2.2.8)). On recalling (5.8), the proof of the lemma is rapidly completed. □

At this point we owe the reader a comment concerning the computations implicit in the proof of Lemma 2.2.2 of [9]. The latter makes fundamental use of the work of Nečaev and Topunov [18], which itself makes extensive use of computers in bounding complete exponential sums over general quartic polynomials. However, in the present context we require such bounds only for the quartic polynomials of the shape bx^4, and, moreover, it suffices to consider a set of coefficients b providing a set of representatives of the cosets modulo fourth powers. In any case, such computations as are implicit in [18] will be easily dispatched in the present setting by an energetic reader equipped with a hand-held calculator. The additional work required to establish the estimates provided in the proof of Lemma 2.2.2 of [9] will be similarly accommodated, since the primes 5, 13, 17, 29, 37, 41, 53, 61, 73 may be directly examined according to the above comments.

The conclusion of Lemma 5.1 is readily applied to bound $G_\varepsilon(q,a)$ via standard methods, although some attention must be paid to the prime 2.

LEMMA 5.2. — *Whenever* $(q, a) = 1$, *one has*
$$|G_1(q,a)| \leqslant |G_0(q,a)| \leqslant q\kappa(q).$$

Proof. — The standard theory of complete exponential sums (see, for example, the proof of Lemma 2.10 of Vaughan [**21**]) shows that the exponential sum $G_\varepsilon(q, a)$ has the quasi-multiplicative property to the effect that, whenever $(q_1, q_2) = 1$ and $(a_i, q_i) = 1$ ($i = 1, 2$), then one has
$$G_\varepsilon(q_1 q_2, a_1 q_2 + a_2 q_1) = G_\varepsilon(q_1, a_1) G_\varepsilon(q_2, a_2). \tag{5.9}$$

Consequently, in view of (5.3) and Lemma 5.1, the conclusion of the lemma will be established by verifying that whenever a is odd and $l \geqslant 1$, one has
$$|G_1(2^l, a)| \leqslant |G_0(2^l, a)| \leqslant 2^l \kappa(2^l). \tag{5.10}$$

Note first that when $\varepsilon \in \{0, 1\}$, one has
$$(2h + \varepsilon)^4 \equiv \varepsilon \pmod{16}, \tag{5.11}$$
and thus we find that when $1 \leqslant l \leqslant 4$, one has
$$|G_\varepsilon(2^l, a)| = 2^l = 2^l \kappa(2^l),$$
thereby confirming (5.10) for $1 \leqslant l \leqslant 4$. Suppose next that $l \geqslant 5$ and that a is odd. In these circumstances, Lemma 4.4 of [**21**] asserts that $S(2^l, a) = 2^3 S(2^{l-4}, a)$, whence
$$G_0(2^l, a) = 2^4 S(2^{l-4}, a) = 2S(2^l, a). \tag{5.12}$$

Since $G_0(2^l, a) + G_1(2^l, a) = 2S(2^l, a)$, it is immediate from (5.12) that whenever $l \geqslant 5$ and a is odd, one has
$$G_1(2^l, a) = 0. \tag{5.13}$$

Meanwhile, on writing $l = 4u + v$ with $u \geqslant 1$ and $1 \leqslant v \leqslant 4$, it follows from (5.12) and Lemma 4.4 of [**21**] that
$$G_0(2^l, a) = 2S(2^{4u+v}, a) = 2^{3u+1} S(2^v, a).$$

But by (5.11) we have
$$|S(2^v, a)| = 2^{v-1} |1 + e(a2^{-v})| \leqslant 2^{v-1} |1 + e(2^{-v})|,$$
and moreover a simple calculation reveals that for $1 \leqslant v \leqslant 4$, one has
$$|1 + e(2^{-v})| = \sqrt{2(1 + \cos(\pi 2^{1-v}))} = c(2^v).$$

Thus we deduce that whenever a is odd and $l \geqslant 5$, one has
$$|G_0(2^l, a)| \leqslant 2^{3u+v} c(2^v) = 2^l \kappa(2^l). \tag{5.14}$$

On combining (5.13) and (5.14), we verify (5.10) for $l \geqslant 5$, and in view of our earlier discussion, the inequalities (5.10) therefore hold for each $l \geqslant 1$. This completes the proof of the lemma. □

We next record a couple of simple lemmata of considerable utility. When p is an odd prime number, define the integer b_p by

$$b_p = (p-1, 4) - 1 = \begin{cases} 1, & \text{when } p \equiv 3 \pmod{4}, \\ 3, & \text{when } p \equiv 1 \pmod{4}. \end{cases} \quad (5.15)$$

LEMMA 5.3. — *When p is an odd prime, one has*

$$\sum_{a=1}^{p-1} |S(p,a)|^2 = b_p p(p-1).$$

Proof. — The sum in question is equal to

$$\sum_{a=1}^{p} |S(p,a)|^2 - p^2 = p \sum_{\substack{1 \leqslant x, y \leqslant p \\ x^4 \equiv y^4 \pmod{p}}} 1 - p^2. \quad (5.16)$$

But for any fixed x with $1 \leqslant x < p$, there are precisely $(p-1, 4)$ values of y with $x^4 \equiv y^4 \pmod{p}$ and $1 \leqslant y \leqslant p$. Thus the final sum in (5.16) is equal to $1 + (p-1, 4)(p-1)$, and the desired conclusion follows immediately from (5.15). □

LEMMA 5.4. — *Let $X \geqslant 5$ be an integer and let $\sigma > 1$ be a real number. Then one has*

$$\sum_{\substack{n \geqslant X \\ n \equiv X \pmod{4}}} n^{-\sigma} < \frac{(X-4)^{1-\sigma}}{4(\sigma-1)}.$$

Proof. — We merely note that $(4z + X)^{-\sigma}$ is a decreasing function of z for $z \geqslant -1$, and hence

$$\sum_{m=0}^{\infty} (4m + X)^{-\sigma} < \int_{-1}^{\infty} (4z + X)^{-\sigma} dz = \frac{(X-4)^{1-\sigma}}{4(\sigma-1)}.$$

This completes the proof of the lemma. □

Next, we consider a sum which plays a role in our evaluation of the major arc contribution $R(N; \mathfrak{M})$. In this context, we write

$$V(q) = q^{-6} \sum_{\substack{a=1 \\ (a,q)=1}}^{q} |G_0(q,a)|^6. \quad (5.17)$$

LEMMA 5.5. — *When $X \geqslant 10^{25}$, one has*

$$\sum_{1 \leqslant q \leqslant X} q^{1/4} V(q) < 1.29 \times 10^6 \log X.$$

Proof. — In view of the relation (5.9), a routine argument (see, for example, the proof of Lemma 2.11 of Vaughan [**21**]) confirms that $V(q)$ is a multiplicative function of q. It therefore follows that

$$\sum_{1 \leqslant q \leqslant X} q^{1/4} V(q) \leqslant \prod_{p \leqslant X} W(p), \qquad (5.18)$$

where we write

$$W(p) = \sum_{l=0}^{\infty} p^{l/4} V(p^l).$$

Observe first that in view of Lemma 5.2, one has

$$p^{l/4} V(p^l) \leqslant \kappa(p^l)^6 p^{5l/4 - 1} (p-1). \qquad (5.19)$$

In particular, it follows from (5.4) that

$$W(2) \leqslant 1 + \sum_{l=1}^{4} 2^{5l/4 - 1} \kappa(2^l)^6 + \sum_{u=1}^{\infty} \sum_{v=1}^{4} 2^{\frac{5}{4}(4u+v) - 1} \kappa(2^{4u+v})^6$$

$$= 1 + \sum_{l=1}^{4} 2^{5l/4 - 1} + \sum_{u=1}^{\infty} 2^{-u-1} \sum_{v=1}^{4} 2^{5v/4} c(2^v)^6,$$

and thus a simple computation yields the estimate

$$W(2) < 28 + 2404 \sum_{u=1}^{\infty} 2^{-u-1} = 1230. \qquad (5.20)$$

We now turn our attention towards the odd primes p. On recalling (5.3) and (5.8), it follows from Lemma 5.1 that for $u \geqslant 0$, one has

$$V(p^{4u+1}) = p^{-6(4u+1)} \sum_{\substack{a=1 \\ (a,p)=1}}^{p^{4u+1}} p^{18u} |S(p,a)|^6$$

$$= p^{-2u-6} \sum_{a=1}^{p-1} |S(p,a)|^6 \leqslant p^{-2u-2} \kappa(p)^4 \sum_{a=1}^{p-1} |S(p,a)|^2.$$

Consequently, we deduce from Lemma 5.3 that

$$p^{(4u+1)/4} V(p^{4u+1}) \leqslant b_p (p-1) p^{-u-3/4} \kappa(p)^4. \qquad (5.21)$$

When $u \geqslant 0$ and $2 \leqslant v \leqslant 4$, meanwhile, we deduce from (5.19) and (5.6) that

$$p^{(4u+v)/4} V(p^{4u+v}) \leqslant (p-1) p^{-u+5v/4 - 7}. \qquad (5.22)$$

On collecting together (5.21) and (5.22), we deduce that

$$W(p) = 1 + \sum_{u=0}^{\infty}\sum_{v=1}^{4} p^{(4u+v)/4} V(p^{4u+v})$$

$$\leqslant 1 + (p-1)\sum_{u=0}^{\infty} p^{-u}\left(b_p p^{-3/4}\kappa(p)^4 + \sum_{v=2}^{4} p^{5v/4-7}\right)$$

$$= 1 + b_p p^{1/4}\kappa(p)^4 + p^{-7/2} + p^{-9/4} + p^{-1}. \quad (5.23)$$

On recalling (5.5) and (5.15), therefore, a modest computation reveals that

$$W(3) < 1.586 \quad \text{and} \quad W(5) < 3.955. \quad (5.24)$$

On the other hand, again by (5.5) and (5.15), for $p > 5$ one has $b_p p^{1/4}\kappa(p)^4 \geqslant p^{-7/4}$, and consequently,

$$\frac{1}{2}(b_p p^{1/4}\kappa(p)^4 + p^{-1})^2 - (p^{-7/2} + p^{-9/4})$$

$$\geqslant \frac{1}{2}(p^{-7/2} + 2p^{-11/4} + p^{-2}) - (p^{-7/2} + p^{-9/4})$$

$$= \frac{1}{2}p^{-2}(1 - p^{-1/2})((1 - p^{-1/4})^2 + p^{-1}) > 0.$$

In this way, the inequality (5.23) leads to the upper bound

$$W(p) < 1 + b_p p^{1/4}\kappa(p)^4 + p^{-1} + \frac{1}{2}(b_p p^{1/4}\kappa(p)^4 + p^{-1})^2$$

$$< \exp(b_p p^{1/4}\kappa(p)^4 + p^{-1}). \quad (5.25)$$

Again recalling (5.5) and (5.15), applying Lemma 5.4, and making a modest computation, we discover that

$$\sum_{\substack{p \geqslant 7 \\ p \equiv 3 \,(\text{mod } 4)}} b_p p^{1/4}\kappa(p)^4 < \sum_{\substack{n \geqslant 7 \\ n \equiv 3 \,(\text{mod } 4)}} n^{-7/4} < 3^{-3/4}/3 < 0.1463,$$

and similarly,

$$\sum_{\substack{p \geqslant 7 \\ p \equiv 1 \,(\text{mod } 4)}} b_p p^{1/4}\kappa(p)^4 = 3\sum_{\substack{13 \leqslant p \leqslant 73 \\ p \equiv 1 \,(\text{mod } 4)}} c(p)^4 p^{-3/4} + 3^5 \sum_{\substack{p \geqslant 89 \\ p \equiv 1 \,(\text{mod } 4)}} p^{-7/4}$$

$$< 2.8295 + 3^5(85^{-3/4}/3) < 5.723.$$

Moreover, when $X \geqslant 10^{25}$ it follows from Lemma 3.2 that

$$\sum_{7 \leqslant p \leqslant X} \frac{1}{p} < \log\log X + 0.281 - \frac{1}{2} - \frac{1}{3} - \frac{1}{5}$$

$$< \log\log X - 0.752.$$

We therefore deduce from (5.25) that

$$\prod_{7\leqslant p\leqslant X} W(p) < \exp\left(\sum_{7\leqslant p\leqslant X} (b_p p^{1/4}\kappa(p)^4 + p^{-1})\right)$$
$$< \exp(0.1463 + 5.723 - 0.752)\log X$$
$$< 167 \log X. \tag{5.26}$$

The proof of the lemma is completed on combining (5.18), (5.20), (5.24) and (5.26). □

Finally, we take this occasion to discuss an estimate for the complete exponential sum

$$S(q, a, b) = \sum_{r=1}^{q} e((ar^4 + br)/q). \tag{5.27}$$

We implicitly require the following result, which was first proved by Thomas in his thesis (see Theorem 2.1 of [20]).

LEMMA 5.6. — *Whenever $(q, a, b) = 1$, one has*

$$|S(q, a, b)| \leqslant 4.5 q^{3/4}.$$

Some comments are in order, before we launch our proof of this lemma. In §7 below, we require Proposition 2.4 of Deshouillers and Dress [9], but the latter proposition is in fact deduced from the aforementioned result of Thomas via Proposition 2.2 of [9]. Since the argument of Nečaev and Topunov [18] is employed in the proof of this result of Thomas by Deshouillers and Dress [9], one finds that the proof of Proposition 2.4 of [9] ultimately rests, implicitly, on the extensive computations within the work of Nečaev and Topunov [18]. Our objective here is to substantially reduce the computational load required to confirm the above bound for $S(q, a, b)$, and hence to establish Proposition 2.4 of Deshouillers and Dress [9]. To this end, we provide here a proof of this result of Thomas [20].

Proof of Lemma 5.6. — By the standard theory of complete exponential sums (see, for example, Lemma 2.10 of Vaughan [21]), one finds easily that whenever $(q_1, q_2) = 1$, then

$$S(q_1 q_2, a_1 q_2 + a_2 q_1, b_1 q_2 + b_2 q_1) = S(q_1, a_1, b_1) S(q_2, a_2, b_2). \tag{5.28}$$

Moreover, under the same condition it follows that

$$(q_1 q_2, a_1 q_2 + a_2 q_1, b_1 q_2 + b_2 q_1) = 1$$

if and only if $(q_1, a_1, b_1) = (q_2, a_2, b_2) = 1$. We therefore find that, in order to establish the conclusion of the lemma, it suffices to consider the sums $S(q, a, b)$ for prime powers $q = p^h$ and integers a and b with $p \nmid (a, b)$.

Observe next that whenever h and ν are integers with $1 \leqslant \nu \leqslant h/2$, one has
$$a(x + yp^{h-\nu})^4 + b(x + yp^{h-\nu}) \equiv ax^4 + bx + (4ax^3 + b)yp^{h-\nu} \pmod{p^h}.$$
When $p = 2$ or 3, moreover, this congruence remains valid under the slightly weaker constraint $1 \leqslant \nu \leqslant (h+1)/2$ and $h \geqslant 2$. Suppose that $q = p^h$ with $h \geqslant 2$, and make the change of variable $r = x + yp^{h-\nu}$ in (5.27), where $1 \leqslant x \leqslant p^{h-\nu}$ and $1 \leqslant y \leqslant p^\nu$. Then one readily deduces from the above congruence that when $1 \leqslant \nu \leqslant h/2$, or when $1 \leqslant \nu \leqslant (h+1)/2$ in the cases $p = 2, 3$, one has
$$S(p^h, a, b) = p^\nu \sum_{\substack{x=1 \\ 4ax^3 \equiv -b \pmod{p^\nu}}}^{p^{h-\nu}} e(p^{-h}(ax^4 + bx)). \tag{5.29}$$

We begin by considering the even prime $p = 2$. When $1 \leqslant h \leqslant 3$, the trivial bound $|S(2^h, a, b)| \leqslant 2^h$ suffices for our purposes. When $h = 4$, we find from (5.11) that
$$S(2^4, a, b) = \sum_{\varepsilon=0}^{1} \sum_{r=1}^{8} e(2^{-4}(a(2r + \varepsilon)^4 + b(2r + \varepsilon)))$$
$$= (1 + e((a + b)/16)) \sum_{r=1}^{8} e(br/8),$$
and from this it plainly follows that whenever $2 \nmid (a, b)$, one has
$$|S(2^4, a, b)| \leqslant |1 + e(1/16)| \cdot 8 = \sqrt{2 + \sqrt{2 + \sqrt{2}}} \cdot 8. \tag{5.30}$$
When $h \geqslant 5$, we first apply the formula (5.29) with $p = 2$ and $\nu = 3$. Note that when $4 \nmid b$, the congruence $4ax^3 \equiv -b \pmod 8$ has no solution, and that when $8 | b$ and $2 \nmid a$, this congruence holds if and only if $2 | x$. Therefore it follows immediately that
$$S(2^h, a, b) = \begin{cases} 0, & \text{when } h \geqslant 5 \text{ and } 4 \nmid b, \\ 8S(2^{h-4}, a, b/8), & \text{when } h \geqslant 5,\ 2 \nmid a \text{ and } 8|b. \end{cases} \tag{5.31}$$
When $h \geqslant 5$, $2 \nmid a$, $4 | b$ but $8 \nmid b$, we apply (5.29) with $p = 2$ and $\nu = [(h+1)/2]$. Then since $h - \nu \geqslant \nu - 2$ and the congruence $ax^3 \equiv -b/4 \pmod{2^{\nu-2}}$ possesses a unique solution x modulo $2^{\nu-2}$ under the current hypotheses, one obtains the estimate
$$|S(2^h, a, b)| \leqslant 2^\nu (2^{h-\nu}/2^{\nu-2}) = 2^{h-[(h+1)/2]+2} \leqslant \sqrt{2} \cdot 2^{3h/4}. \tag{5.32}$$
On collecting together the relations (5.30)-(5.32), and noting also the trivial bound for $1 \leqslant h \leqslant 3$, an inductive argument demonstrates that whenever $2 \nmid (a, b)$ and $h \geqslant 1$, one has
$$|S(2^h, a, b)| \leqslant \sqrt{2 + \sqrt{2 + \sqrt{2}}} \cdot 2^{3h/4}. \tag{5.33}$$
We hereafter concentrate on odd primes p. Suppose first that $h \geqslant 2$, $p \nmid a$ and $p | b$. In this case we take $\nu = 1$ in (5.29), and note that the congruence $4ax^3 \equiv -b$ (mod p) is equivalent to the condition that $p | x$. When $h \leqslant 4$, the latter observation

ensures that the sum on the right hand side of (5.29) is easily evaluated, and thus one obtains

$$S(p^h, a, b) = \begin{cases} p^{h-1}, & \text{when } 2 \leqslant h \leqslant 4 \text{ and } p^{h-1}|b, \\ 0, & \text{when } 3 \leqslant h \leqslant 4, p|b \text{ but } p^{h-1} \nmid b. \end{cases} \quad (5.34)$$

When $h \geqslant 5$, meanwhile, we may make the change of variable $x = p(z + wp^{h-4})$ in (5.29), with $1 \leqslant z \leqslant p^{h-4}$ and $1 \leqslant w \leqslant p^2$. In this way, we deduce that

$$S(p^h, a, b) = \begin{cases} p^3 S(p^{h-4}, a, b/p^3), & \text{when } h \geqslant 5 \text{ and } p^3|b, \\ 0, & \text{when } h \geqslant 5, p|b \text{ but } p^3 \nmid b. \end{cases} \quad (5.35)$$

Suppose next that $h \geqslant 2$ and $p \nmid b$. Write $\theta(p, u, v)$ for the number of solutions of the congruence $4ax^3 \equiv -b \pmod{p^u}$ with $1 \leqslant x \leqslant p^v$. Since $p \nmid b$, one finds for every natural number u that $\theta(p, u, u) \leqslant (p^{u-1}(p-1), 3)$, and also that $\theta(3, u, u-1) \leqslant 1$. On taking $\nu = [h/2]$ when $p > 3$, and $\nu = [(h+1)/2]$ when $p = 3$, we therefore derive from (5.29) the bound

$$|S(p^h, a, b)| \leqslant \theta(p, \nu, h - \nu) p^\nu.$$

A modicum of computation therefore confirms that when $h \geqslant 2$, one has

$$|S(3^h, a, b)| \leqslant 3^{3h/4}, \quad (5.36)$$

and likewise that when h is even and $p > 3$,

$$|S(p^h, a, b)| \leqslant (p-1, 3) p^{h/2}. \quad (5.37)$$

When $p > 3$ and h is odd, moreover, one has $h = 2\nu + 1$, and on writing $x = z + wp^\nu$ in (5.29), one deduces that

$$|S(p^h, a, b)| \leqslant p^\nu \sum_{\substack{z=1 \\ 4az^3 \equiv -b \pmod{p^\nu}}}^{p^\nu} \left| \sum_{w=1}^{p} e\big((6az^2 w^2 + (4az^3 + b)wp^{-\nu})p^{-1}\big) \right|. \quad (5.38)$$

Since, by hypothesis, we may suppose that $p \nmid 6b$, we find that whenever an integer z satisfies the congruence appearing in the summation condition on the right hand side of the last inequality, one necessarily has $p \nmid 6az^2$. On considering the associated Gauss sums, we therefore deduce from (5.38) that $|S(p^h, a, b)| \leqslant \theta(p, \nu, \nu) p^{\nu + 1/2}$, and thus the inequality (5.37) remains valid for $p > 3$ and odd exponents h exceeding 2. On recalling the upper bounds (5.36) and (5.37), we thus conclude that when p is odd, $p \nmid b$, $p \neq 7$ and $h \geqslant 2$, or when $p = 7$, $7 \nmid b$ and $h \geqslant 3$, one has the bound

$$|S(p^h, a, b)| \leqslant p^{3h/4}. \quad (5.39)$$

Before proceeding further, we remark that by a transparent change of variable, and by considering complex conjugation, one finds that whenever r is an integer with $(q, r) = 1$, then

$$|S(q, a, b)| = |S(q, \pm ar^4, br)|. \quad (5.40)$$

Equipped with the relation (5.40), we begin by disposing of the case in which $p = 7$, $7 \nmid b$ and $h = 2$. Here we appeal to (5.29) with $\nu = 1$ just as before. Since cubic residues are congruent to 0 or ± 1 modulo 7, we find that the congruence condition in (5.29) ensures that the sum $S(49, a, b)$ vanishes unless $4a \equiv \pm b \pmod{7}$. In view of (5.40), moreover, on noting that the integers $\pm r^4$ with $7 \nmid r$ represent all the reduced residue classes modulo 49, it suffices to consider only the cases in which $a = 1$ and $b \equiv 3 \pmod{7}$ with $|b| \leq 24$. Indeed, a modicum of computation reveals that

$$\max_{\substack{|b| \leq 24 \\ b \equiv 3 \pmod 7}} \frac{1}{\sqrt{7}} \left| \sum_{x \in \{1,2,-3\}} e\left(\frac{x^4 + bx}{49}\right) \right| < 1.035.$$

It may be worth noting here that the maximum on the left hand side occurs when $b = -18$. In any case, it follows from (5.29), combined with the above observations, that whenever $7 \nmid b$, one has

$$|S(49, a, b)| < 1.035 \cdot 49^{3/4}. \tag{5.41}$$

It remains only to estimate $S(p, a, b)$ for odd primes p with $p \nmid (a, b)$. But when $p | a$ the sum $S(p, a, b)$ vanishes, and when $p | b$ this sum is identical with the Gauss sum $S(p, a)$ already estimated in the course of the proof of Lemma 5.1. In any case, when $p \nmid (a, b)$ and $p | ab$, it follows that

$$|S(p, a, b)| \leq c(p) p^{3/4}, \tag{5.42}$$

where $c(p)$ is defined as in the preamble to (5.4) and (5.5) above.

Our subsequent treatment for the case with $p \nmid ab$ is motivated by the method due to Mordell (see the proof of Theorem 7.1 of Vaughan [21], for example). We consider the sums

$$\Upsilon_0 = \sum_{a=0}^{p-1} \sum_{b=0}^{p-1} |S(p, a, b)|^4 \quad \text{and} \quad \Upsilon_1 = \sum_{a=1}^{p-1} \sum_{b=1}^{p-1} |S(p, a, b)|^4.$$

In order to derive a relation between these sums, we begin by applying Cauchy's inequality in combination with Lemma 5.3 to obtain the lower bound

$$\sum_{a=1}^{p-1} |S(p, a, 0)|^4 \geq \left(\sum_{a=1}^{p-1} |S(p, a)|^2 \right)^2 \left(\sum_{a=1}^{p-1} 1 \right)^{-1} \geq p^2 (p-1).$$

Observing next that $S(p, 0, b)$ is p or 0 according to whether $p | b$ or not, we deduce that

$$\Upsilon_1 \leq \Upsilon_0 - p^4 - p^2(p-1). \tag{5.43}$$

On the other hand, it follows from (5.40) that whenever $p \nmid ab$, one has

$$\Upsilon_1 \geq 2(p-1)|S(p, a, b)|^4. \tag{5.44}$$

We thus conclude from (5.43) and (5.44) that whenever $p \nmid ab$, one has
$$|S(p,a,b)| \leqslant \left(\frac{\Upsilon_0 - p^4 - p^3 + p^2}{2(p-1)}\right)^{1/4}, \qquad (5.45)$$
and it is this relation that we employ in what follows.

We note next that orthogonality yields the equation
$$\Upsilon_0 = p^2 \Upsilon_2, \qquad (5.46)$$
where we write Υ_2 for the number of solutions of the simultaneous congruences
$$x^4 + y^4 \equiv z^4 + w^4 \pmod{p} \quad \text{and} \quad x + y \equiv z + w \pmod{p},$$
subject to $1 \leqslant x, y, z, w \leqslant p$. Eliminating w from these congruences, one finds that Υ_2 is equal to the number of solutions of the congruence
$$(x-z)(y-z)((4x+3y-z)^2 + 7y^2 - 2yz + 7z^2) \equiv 0 \pmod{p},$$
subject to $1 \leqslant x, y, z \leqslant p$. The number of these solutions with $x = z$ or $y = z$ is $2p^2 - p$, and thus, on denoting the Legendre symbol modulo p by χ_p, one obtains the upper bound
$$\Upsilon_2 \leqslant 2p^2 - p + \sum_{\substack{1 \leqslant y, z \leqslant p \\ y \neq z}} \left(1 + \chi_p(-7y^2 + 2yz - 7z^2)\right). \qquad (5.47)$$
Familiar results for sums of Legendre symbols over quadratic sequences demonstrate that for each prime p exceeding 3, one has
$$\sum_{y=1}^{p} \sum_{z=1}^{p} \chi_p(-7y^2 + 2yz - 7z^2) = 0.$$
Then it follows swiftly from (5.47) that for $p > 3$, one has
$$\Upsilon_2 \leqslant 3p^2 - 2p - \chi_p(-3)(p-1) \leqslant 3p^2 - p - 1. \qquad (5.48)$$

We now collect together (5.45), (5.46) and (5.48) to conclude that whenever $p > 3$ and $p \nmid ab$, one has
$$|S(p,a,b)| \leqslant p^{3/4}.$$
But when $p = 3$ and $3 \nmid ab$, one has
$$|S(3,a,b)| = |1 + e((a+b)/3) + e((a-b)/3)| = |2 + e(2/3)| = \sqrt{3}.$$
We therefore conclude that the inequality (5.42) is valid whenever p is an odd prime number with $p \nmid (a,b)$.

Define now $c^*(2) = c(2^4)$, $c^*(7) = 1.035$, and when p is an odd prime number with $p \neq 7$, define $c^*(p) = c(p)$. Then in view of (5.34), (5.35), (5.39), (5.41) and the conclusion of the previous paragraph, we deduce via an inductive argument that whenever p is odd, $p \nmid (a,b)$ and $h \geqslant 1$, one has
$$|S(p^h, a, b)| \leqslant c^*(p) p^{3h/4}.$$

Finally, on recalling (5.28) and (5.33), the proof of the lemma is completed by observing that
$$c^*(2)c^*(5)c^*(7)c^*(13)c^*(17)c^*(41) < 4.42.$$
□

CHAPTER 6

THE SINGULAR SERIES

In the analysis of the major arc contribution described in the next section, we encounter a partial sum of the singular series. The object of the present section is to obtain a uniform lower bound for this partial sum adequate for the aforementioned application. We begin by recording an elementary fact concerning fourth power residues in a form suitable for frequent citation in our later discussion.

LEMMA 6.1. — *When p is a prime number, define $\gamma = \gamma(p)$ by*

$$\gamma(p) = \begin{cases} 4, & \text{when } p = 2, \\ 1, & \text{otherwise.} \end{cases}$$

Suppose that s is a natural number, and that $\phi(\boldsymbol{x}) \in \mathbb{Z}[x_1, \ldots, x_s]$. Let $a_1, \ldots, a_s$ be fixed integers, and write $m(\boldsymbol{a})$ for the number of solutions of the congruence

$$x^4 \equiv \phi(a_1, \ldots, a_s) \pmod{p^\gamma},$$

with $1 \leqslant x \leqslant p^\gamma$ and $p \nmid x$. When l is a natural number, denote by $\mathcal{M}(p^l; \boldsymbol{a})$ the number of solutions of the congruence

$$x^4 \equiv \phi(x_1, \ldots, x_s) \pmod{p^l},$$

with $1 \leqslant x \leqslant p^l$, $p \nmid x$ and

$$x_j \equiv a_j \pmod{p^\gamma}, \quad 1 \leqslant x_j \leqslant p^l \quad (1 \leqslant j \leqslant s). \tag{6.1}$$

Then whenever $l \geqslant \gamma$, one has

$$\mathcal{M}(p^l; \boldsymbol{a}) = p^{s(l-\gamma)} m(\boldsymbol{a}).$$

Proof. — When $m(\boldsymbol{a})$ is zero, the conclusion of the lemma is trivial, so we suppose henceforth that $m(\boldsymbol{a})$ is non-zero. The theory of primitive roots (see, for example, Chapter 10 of Apostol [1]) shows that when p is an odd prime with $p \nmid n$, and n is a fourth power residue modulo p, then for every $l \geqslant 1$ the congruence $x^4 \equiv n \pmod{p^l}$ has precisely $(p-1, 4)$ solutions distinct modulo p^l. When $p = 2$, meanwhile, the condition $2 \nmid x$ ensures that $x^4 \equiv 1 \pmod{16}$, and we see that it is only the congruence class 1 modulo 16 that is relevant. But when $n \equiv 1 \pmod{16}$, it follows from Theorem 10.11 of Apostol [**1**], for example, that for every $l \geqslant 4$ the congruence $x^4 \equiv n \pmod{2^l}$ has precisely 8 solutions distinct modulo 2^l. Then in either case one finds that whenever $m(\boldsymbol{a}) > 0$, one has $m(\boldsymbol{a}) = 8$ when $p = 2$, and $m(\boldsymbol{a}) = (p-1, 4)$ when $p > 2$. The proof of the lemma is therefore completed by noting that the number of s-tuples $\boldsymbol{x}$ satisfying (6.1) is $p^{s(l-\gamma)}$. □

We are now equipped to establish the desired lower bound for a partial sum of the singular series.

LEMMA 6.2. — *Let t be an integer with $1 \leqslant t \leqslant 7$, and suppose that $n \equiv t \pmod{16}$. Define $A(q, n; t)$ and $\mathfrak{S}(n, Q; t)$ by*

$$A(q, n; t) = q^{-7} \sum_{\substack{a=1 \\ (a,q)=1}}^{q} G_0(q,a)^{7-t} G_1(q,a)^t e(-an/q),$$

and

$$\mathfrak{S}(n, Q; t) = \sum_{q \leqslant Q} A(q, n; t).$$

Then whenever $P \geqslant 10^{50}$, one has

$$\mathfrak{S}(n, P^{1/2}; t) > 1.269 \qquad \text{when } n \equiv 3 \pmod{5},$$

and

$$\mathfrak{S}(n, P^{1/2}; t) > 5.078 \qquad \text{when } n \not\equiv 3 \pmod{5}.$$

Proof. — We begin by showing that the truncated singular series $\mathfrak{S}(n, Q; t)$ is close to the corresponding infinite sum. On recalling the definition (5.17) of $V(q)$, we deduce from (5.7) and Lemma 5.2 that

$$|A(q, n; t)| \leqslant 9 q^{-1/4} V(q).$$

Consequently, whenever $Y \geqslant P^{1/2}$, it follows via partial summation that

$$\sum_{P^{1/2} < q \leqslant Y} |A(q, n; t)| \leqslant 9 \sum_{P^{1/2} < q \leqslant Y} q^{-1/4} V(q)$$

$$= 9 Y^{-1/2} \sum_{1 \leqslant q \leqslant Y} q^{1/4} V(q) - 9 P^{-1/4} \sum_{1 \leqslant q \leqslant P^{1/2}} q^{1/4} V(q)$$

$$+ \frac{9}{2} \int_{P^{1/2}}^{Y} X^{-3/2} \sum_{1 \leqslant q \leqslant X} q^{1/4} V(q) \, dX.$$

On recalling the conclusion of Lemma 5.5, we therefore deduce that whenever $P \geqslant 10^{50}$ one has

$$\sum_{P^{1/2}<q\leqslant Y} |A(q,n;t)| < 9 \times 1.29 \times 10^6 \left(Y^{-1/2} \log Y + \frac{1}{2} \int_{P^{1/2}}^{Y} X^{-3/2} \log X \, dX \right)$$

$$< 1.16 \times 10^7 \left(P^{-1/4} \log(P^{1/2}) + 2P^{-1/4} \right)$$

$$< 3 \times 10^{-4}.$$

We thus deduce that the infinite series $\mathfrak{S}(n;t)$, defined by

$$\mathfrak{S}(n;t) = \sum_{q=1}^{\infty} A(q,n;t),$$

converges absolutely, and moreover that whenever $P \geqslant 10^{50}$, one has

$$|\mathfrak{S}(n;t) - \mathfrak{S}(n, P^{1/2};t)| < 3 \times 10^{-4}. \tag{6.2}$$

We next express the singular series $\mathfrak{S}(n;t)$ as an absolutely convergent infinite product amenable to our subsequent discussion. Denote by $M(q,n;t)$ the number of solutions of the congruence

$$\sum_{j=1}^{7-t} (2x_j)^4 + \sum_{l=1}^{t} (2y_l+1)^4 \equiv n \pmod{q}, \tag{6.3}$$

with $1 \leqslant x_j \leqslant q$ ($1 \leqslant j \leqslant 7-t$) and $1 \leqslant y_l \leqslant q$ ($1 \leqslant l \leqslant t$). The standard theory of complete exponential sums (see, for example, Lemma 2.12 of Vaughan [21] or §2.3 of Deshouillers and Dress [9]) shows that

$$M(q,n;t) = q^{-1} \sum_{a=1}^{q} G_0(q,a)^{7-t} G_1(q,a)^t e\left(-\frac{a}{q}n\right)$$

$$= q^6 \sum_{d|q} A(d,n;t). \tag{6.4}$$

But $M(q,n;t)$ is plainly a multiplicative function of q, and thus we deduce that $A(q,n;t)$ is likewise a multiplicative function of q. Consequently, on defining $B(p,n;t)$ by

$$B(p,n;t) = \sum_{l=0}^{\infty} A(p^l,n;t),$$

we may express $\mathfrak{S}(n;t)$ as

$$\mathfrak{S}(n;t) = \prod_p B(p,n;t). \tag{6.5}$$

Here we note that the absolute convergence of the latter infinite product is assured by that of the series $\sum_q |A(q,n;t)|$.

We must now estimate the local factors $B(p,n;t)$ in the infinite product (6.5). While for larger p this estimation is essentially routine, we must obtain estimates for smaller p of some precision, and this entails some moderately painful computation. We first consider the factor $B(2, n; t)$. Since we are assuming that $n \equiv t \pmod{16}$, we deduce from (5.11) that when $q = 16$ the congruence (6.3) holds for every choice of $\boldsymbol{x}$ and $\boldsymbol{y}$, whence we find that $M(2^4, n; t) = 2^{28}$. On recalling (5.13) and (6.4), therefore, we arrive at the relation

$$B(2, n; t) = \sum_{l=0}^{4} A(2^l, n; t) = 2^{-24} M(2^4, n; t) = 16. \qquad (6.6)$$

In order to estimate $B(p, n; t)$ for odd primes p, we work with the number of solutions of an auxiliary congruence. Denote by $M_1(q, n)$ the number of solutions of the congruence

$$x_1^4 + x_2^4 + \cdots + x_7^4 \equiv n \pmod{q},$$

with $1 \leqslant x_j \leqslant q$ ($1 \leqslant j \leqslant 7$), and let $M_1^*(q, n)$ denote the corresponding number of solutions subject to the additional condition that $(x_j, q) = 1$ for some j. Then, on combining a trivial estimate with the conclusion of Lemma 6.1, one finds that for $l \geqslant 1$ and every odd prime p, one has

$$M_1(p^l, n) \geqslant M_1^*(p^l, n) = p^{6(l-1)} M_1^*(p, n). \qquad (6.7)$$

But it is evident that when q is odd, one has $M(q, n; t) = M_1(q, n)$ for any t with $1 \leqslant t \leqslant 7$, and thus we conclude from (6.4) and (6.7) that for odd primes p, one has

$$B(p, n; t) = \lim_{l \to \infty} p^{-6l} M(p^l, n; t) = \lim_{l \to \infty} p^{-6l} M_1(p^l, n) \geqslant p^{-6} M_1^*(p, n). \qquad (6.8)$$

For each fixed value of p, it is possible to compute $M_1^*(p, n)$ for each value of n modulo p in order to determine the minimal value of $B(p, n; t)$. We begin by noting that when $p = 3$ or 5, one has

$$x^4 \equiv \begin{cases} 0 \pmod{p}, & \text{when } p \mid x, \\ 1 \pmod{p}, & \text{when } p \nmid x, \end{cases}$$

whence it is easily verified that

$$M_1^*(p, n) = \sum_{\substack{1 \leqslant s \leqslant 7 \\ s \equiv n \pmod{p}}} \binom{7}{s} (p-1)^s.$$

By examining this formula for each value of n with $0 \leqslant n < p$, one swiftly verifies that

$$\min_n M_1^*(3, n) = M_1^*(3, 1) = 702,$$

$$\min_n M_1^*(5, n) = M_1^*(5, 3) = 2240,$$

and moreover that

$$\min_{n \not\equiv 3 \pmod{5}} M_1^*(5, n) = M_1^*(5, 4) = 8960.$$

On recalling (6.8), we therefore deduce that
$$B(3,n;t) \geqslant \frac{26}{27}, \qquad B(5,n;t) \geqslant \frac{448}{3125}, \tag{6.9}$$
and when $n \not\equiv 3 \pmod 5$, we find that
$$B(5,n;t) \geqslant \frac{1792}{3125}. \tag{6.10}$$

We next calculate the values $M_1^*(13,n)$ for $0 \leqslant n \leqslant 12$. Observe first that the number $\rho(m)$ of solutions of the congruence $h^4 \equiv m \pmod{13}$, with $0 \leqslant h < 13$, satisfies
$$\rho(m) = \begin{cases} 4, & \text{when } m \equiv 1,\ 3,\ 9 \pmod{13}, \\ 1, & \text{when } m \equiv 0 \pmod{13}, \\ 0, & \text{otherwise.} \end{cases}$$
Thus, on introducing the polynomial
$$f_0(x) = x + x^3 + x^9,$$
we obtain the polynomial congruence
$$(1 + 4f_0(x))^7 \equiv \sum_{n=0}^{12} M_1(13,n) x^n \pmod{x^{13} - 1}. \tag{6.11}$$

In order more easily to compute the left hand side of (6.11), we introduce the auxiliary polynomials
$$f_1(x) = x^2 + x^5 + x^6, \quad f_2(x) = x^4 + x^{10} + x^{12}, \quad f_3(x) = x^7 + x^8 + x^{11},$$
and observe that the following relations hold modulo $x^{13} - 1$:
$$\begin{aligned} &f_0(x)^2 \equiv f_1(x) + 2f_2(x), & &f_0(x)f_1(x) \equiv f_0(x) + f_1(x) + f_3(x), \\ &f_0(x)f_2(x) \equiv f_1(x) + f_3(x) + 3, & &f_0(x)f_3(x) \equiv f_0(x) + f_2(x) + f_3(x). \end{aligned} \tag{6.12}$$

We infer that for $m \geqslant 0$, there are integers $a_j(m)$ ($0 \leqslant j \leqslant 4$) for which
$$f_0(x)^m \equiv a_0(m)f_0(x) + a_1(m)f_1(x) + a_2(m)f_2(x) + a_3(m)f_3(x) + a_4(m)$$
modulo $x^{13} - 1$. Furthermore, in view of (6.12), one has the relation
$$\begin{aligned} f_0(x)^{m+1} \equiv{}& a_0(m)(f_1(x) + 2f_2(x)) + a_1(m)(f_0(x) + f_1(x) + f_3(x)) \\ &+ a_2(m)(f_1(x) + f_3(x) + 3) + a_3(m)(f_0(x) + f_2(x) + f_3(x)) \\ &+ a_4(m)f_0(x) \end{aligned}$$
modulo $x^{13} - 1$, and from this we obtain for $m \geqslant 0$ the recurrence relations
$$\begin{aligned} &a_0(m+1) = a_1(m) + a_3(m) + a_4(m), & &a_1(m+1) = a_0(m) + a_1(m) + a_2(m), \\ &a_2(m+1) = 2a_0(m) + a_3(m), & &a_3(m+1) = a_1(m) + a_2(m) + a_3(m), \\ &a_4(m+1) = 3a_2(m). & & \end{aligned}$$

Since $a_4(0) = 1$ and $a_j(0) = 0$ for $0 \leqslant j \leqslant 3$, one may apply the latter formulae to calculate the values of $a_j(m)$ successively for $m = 1, \ldots, 7$. The following table displays the values of $a_j(m)$ thus obtained for $0 \leqslant m \leqslant 7$ and $0 \leqslant j \leqslant 4$.

m	0	1	2	3	4	5	6	7
$a_0(m)$	0	1	0	1	12	10	51	217
$a_1(m)$	0	0	1	3	4	21	61	147
$a_2(m)$	0	0	2	0	5	30	35	168
$a_3(m)$	0	0	0	3	6	15	66	162
$a_4(m)$	1	0	0	6	0	15	90	105

We calculate the value of $M_1(13, n)$ for $0 \leqslant n \leqslant 12$ by means of (6.11) and the relation

$$(1 + 4f_0(x))^7 = \sum_{m=0}^{7} \binom{7}{m} 4^m f_0(x)^m.$$

In view of our expansion of $f_0(x)^m$ in terms of the $a_j(m)$, we deduce that

$$\sum_{n=0}^{12} M_1(13, n) x^n \equiv b_0 f_0(x) + b_1 f_1(x) + b_2 f_2(x) + b_3 f_3(x) + b_4 \pmod{x^{13} - 1},$$

where for $0 \leqslant j \leqslant 4$ we write

$$b_j = \sum_{m=0}^{7} \binom{7}{m} 4^m a_j(m).$$

We remark that since the left and right hand sides of the last congruence have degree at most 12, then they are in fact equal. We note also that $M_1^*(13, n)$ is equal to $M_1(13, n) - 1$ when $13 | n$, and otherwise is equal to $M_1(13, n)$. Thus, on making use of the table of coefficients $a_j(m)$ presented above, we deduce that

$$\min_n M_1^*(13, n) = \min\{\min_{0 \leqslant j \leqslant 3} b_j, b_4 - 1\} = b_2 = 4\,446\,624.$$

We consequently conclude from (6.8) that

$$B(13, n; t) \geqslant 13^{-6} \times 4\,446\,624 > 0.9212. \tag{6.13}$$

We obtain a lower bound for $B(17, n; t)$ in a similar manner. We write

$$g_0(x) = x + x^4 + x^{13} + x^{16}, \qquad g_1(x) = x^2 + x^8 + x^9 + x^{15},$$
$$g_2(x) = x^3 + x^5 + x^{12} + x^{14}, \qquad g_3(x) = x^6 + x^7 + x^{10} + x^{11},$$

and observe that modulo $x^{17} - 1$, one has

$$(1 + 4g_0(x))^7 \equiv \sum_{n=0}^{16} M_1(17, n) x^n, \tag{6.14}$$

and
$$g_0(x)^2 \equiv g_1(x) + 2g_2(x) + 4, \qquad g_0(x)g_1(x) \equiv g_0(x) + g_1(x) + g_2(x) + g_3(x),$$
$$g_0(x)g_2(x) \equiv 2g_0(x) + g_1(x) + g_3(x), \qquad g_0(x)g_3(x) \equiv g_1(x) + g_2(x) + 2g_3(x).$$

We thus infer on this occasion that for $m \geqslant 0$, there are integers $c_j(m)$ ($0 \leqslant j \leqslant 4$) for which
$$g_0(x)^m \equiv c_0(m)g_0(x) + c_1(m)g_1(x) + c_2(m)g_2(x) + c_3(m)g_3(x) + c_4(m),$$
modulo $x^{17} - 1$, and as before we obtain for $m \geqslant 0$ the recurrence relations
$$c_0(m+1) = c_1(m) + 2c_2(m) + c_4(m), \quad c_1(m+1) = c_0(m) + c_1(m) + c_2(m) + c_3(m),$$
$$c_2(m+1) = 2c_0(m) + c_1(m) + c_3(m), \quad c_3(m+1) = c_1(m) + c_2(m) + 2c_3(m),$$
$$c_4(m+1) = 4c_0(m).$$

Since $c_4(0) = 1$ and $c_j(0) = 0$ for $0 \leqslant j \leqslant 3$, one may apply the latter formulae to calculate the values of $c_j(m)$ successively for $m = 1, \ldots, 7$. The following table displays the values of $c_j(m)$ thus obtained for $0 \leqslant m \leqslant 7$ and $0 \leqslant j \leqslant 4$.

m	0	1	2	3	4	5	6	7
$c_0(m)$	0	1	0	9	5	100	147	1281
$c_1(m)$	0	0	1	3	16	55	251	924
$c_2(m)$	0	0	2	1	24	36	315	756
$c_3(m)$	0	0	0	3	10	60	211	988
$c_4(m)$	1	0	4	0	36	20	400	588

We calculate the value of $M_1(17, n)$ for $0 \leqslant n \leqslant 16$ by means of (6.14) and the relation
$$(1 + 4g_0(x))^7 = \sum_{m=0}^{7} \binom{7}{m} 4^m g_0(x)^m.$$

By employing our expansions for $g_0(x)^m$ in terms of the $c_j(m)$, we deduce that
$$\sum_{n=0}^{16} M_1(17, n)x^n \equiv d_0 g_0(x) + d_1 g_1(x) + d_2 g_2(x) + d_3 g_3(x) + d_4 \pmod{x^{17} - 1},$$

where for $0 \leqslant j \leqslant 4$ we write
$$d_j = \sum_{m=0}^{7} \binom{7}{m} 4^m c_j(m).$$

We note again that $M_1^*(17, n)$ is equal to $M_1(17, n) - 1$ when $17 | n$, and otherwise is equal to $M_1(17, n)$. Thus, on making use of the table of coefficients $c_j(m)$ presented above, we deduce that
$$\min_n M_1^*(17, n) = \min\{\min_{0 \leqslant j \leqslant 3} d_j, d_4 - 1\} = d_4 - 1 = 21\,856\,576.$$

In this way, we conclude from (6.8) that
$$B(17, n; t) \geqslant 17^{-6} \times 21\,856\,576 > 0.9055. \tag{6.15}$$

For the remaining primes p, we apply the lower bound
$$M_1^*(p, n) \geqslant M_1(p, n) - 1, \tag{6.16}$$
in combination with the formula
$$M_1(p, n) = p^{-1} \sum_{a=1}^{p} S(p, a)^7 e(-an/p).$$

On applying Lemmata 5.1 and 5.3 to the latter sum, we obtain
$$M_1(p, n) \geqslant p^6 - p^4 \kappa(p)^5 \sum_{a=1}^{p-1} |S(p, a)|^2$$
$$= p^6 - b_p p^5 (p-1) \kappa(p)^5,$$
whence by (6.8) and (6.16), we deduce that
$$B(p, n; t) \geqslant 1 - b_p(1 - p^{-1})\kappa(p)^5 - p^{-6}. \tag{6.17}$$

When $p \equiv 3 \pmod{4}$, we obtain from (6.17) the lower bound
$$B(p, n; t) \geqslant 1 - (1 - p^{-1})p^{-5/2} - p^{-6}$$
$$\geqslant 1 - p^{-5/2} + \frac{1}{2}p^{-5} > \exp(-p^{-5/2}).$$

Thus, as a consequence of Lemma 5.4, we conclude that
$$\prod_{\substack{p \geqslant 7 \\ p \equiv 3 \pmod{4}}} B(p, n; t) > \exp\left(-\sum_{\substack{p \geqslant 7 \\ p \equiv 3 \pmod{4}}} p^{-5/2}\right)$$
$$\geqslant \exp(-3^{-3/2}/6) > 0.9684. \tag{6.18}$$

When $p \equiv 1 \pmod{4}$ and $p \geqslant 89$, we find from (6.17) that
$$B(p, n; t) \geqslant 1 - 3^6 p^{-5/2}(1 - p^{-1}) - p^{-6} \geqslant 1 - 3^6 p^{-5/2} + \frac{1}{2}3^{12} p^{-5}$$
$$> \exp(-729 p^{-5/2}).$$

In this case, again by Lemma 5.4, we obtain
$$\prod_{\substack{p \geqslant 89 \\ p \equiv 1 \pmod{4}}} B(p, n; t) > \exp\left(-729 \sum_{\substack{p \geqslant 89 \\ p \equiv 1 \pmod{4}}} p^{-5/2}\right)$$
$$> \exp(-729 \times 85^{-3/2}/6) > 0.8563. \tag{6.19}$$

Meanwhile, on recalling the definition (5.5) of $\kappa(p)$, we find from (6.17) that
$$B(p, n; t) \geqslant 1 - 3c(p)^5 p^{-5/4}(1 - p^{-1}) - p^{-6},$$

and thus a modest computation reveals that
$$\prod_{\substack{29 \leqslant p \leqslant 73 \\ p \equiv 1 \pmod 4}} B(p, n; t) > 0.8310. \tag{6.20}$$

Finally, on collecting together (6.6), (6.9), (6.13), (6.15), (6.18), (6.19) and (6.20), and recalling also (6.5), we obtain
$$\mathfrak{S}(n;t) = \prod_p B(p, n; t) > 1.2696.$$

When $n \not\equiv 3 \pmod 5$, moreover, we may substitute (6.10) for the second lower bound of (6.9) in the above calculation, thereby obtaining the stronger lower bound
$$\mathfrak{S}(n;t) > 4 \times 1.2696 = 5.0784.$$

The conclusion of Lemma 6.2 is now immediate on recalling (6.2). □

CHAPTER 7

THE MAJOR ARC CONTRIBUTION

The aim of this section is to evaluate the integral $R(N; \mathfrak{M})$ defined in §2, and thereby to establish Lemma 2.1. We begin by recording an estimate associated with the major arc approximations to the generating functions underlying our methods. Recall the notation of §2, noting especially (2.10), and recall the definition (5.2) of $G_\varepsilon(q,a)$. We define $I(\beta) = I(\beta; N)$ by

$$I(\beta; N) = \int_{2P_0}^{4P_0} e(\beta z^4) dz,$$

and when $\varepsilon \in \{0, 1\}$, we define also

$$T_\varepsilon(q, a, \beta) = (2q)^{-1} G_\varepsilon(q, a) I(\beta).$$

LEMMA 7.1. — *Let α be a real number, and suppose that $a \in \mathbb{Z}$ and $q \in \mathbb{N}$ satisfy the conditions $1 \leqslant a \leqslant q \leqslant P^{1/2}$, $(q, a) = 1$ and $|q\alpha - a| \leqslant 975 P^{-3}$. Write $\beta = \alpha - a/q$. Then whenever $\varepsilon \in \{0, 1\}$, one has*

$$|S_\varepsilon(\alpha) - T_\varepsilon(q, a, \beta)| \leqslant 3 \times 10^6 q^{1/4} P^{1/2}.$$

Proof. — The desired conclusion is immediate from Proposition 2.4 of Deshouillers and Dress [9], but refer to our comment following the statement of our Lemma 5.6 above. □

Equipped with the above estimate, it is essentially routine to exploit our earlier work to derive the auxiliary major arc estimate recorded in the following lemma. We first require some additional notation. Recalling the notation introduced in §2, when $1 \leqslant t \leqslant 7$ we define $\Phi(n; t)$ by

$$\Phi(n; t) = \int_{\mathfrak{M}} S_0(\alpha)^{7-t} S_1(\alpha)^t e(-n\alpha) d\alpha.$$

LEMMA 7.2. — *There exists a real number ν, with $23 < \nu < 64$, for which the following conclusion holds. Suppose that $1 \leqslant t \leqslant 7$, and that n is an integer with*

$$N - 4P^4 \leqslant n \leqslant N \quad \text{and} \quad n \equiv t \pmod{16}.$$

Then whenever $P \geqslant 10^{50}$, one has

$$\Phi(n;t) > 0.000789 P^3 \quad \text{when } n \equiv 3 \pmod 5,$$

and

$$\Phi(n;t) > 0.00316 P^3 \quad \text{when } n \not\equiv 3 \pmod 5.$$

Proof. — Throughout this proof, it is convenient to abbreviate $T_\varepsilon(q, a, \beta)$ as T_ε. Also, when $\alpha = \beta + a/q$, we write

$$U_\varepsilon = S_\varepsilon(\alpha) - T_\varepsilon \quad \text{and} \quad U = 3 \times 10^6 q^{1/4} P^{1/2}.$$

Our first objective is then to compare $\Phi(n;t)$ with the approximation $\Phi_1(n;t)$, which we define by

$$\Phi_1(n;t) = \sum_{q \leqslant P^{1/2}} \sum_{\substack{a=1 \\ (a,q)=1}}^{q} \int_{|\beta| \leqslant 975(qP^3)^{-1}} T_0^{7-t} T_1^t e(-(a/q + \beta)n) d\beta. \tag{7.1}$$

Suppose that a, q and β satisfy $1 \leqslant a \leqslant q \leqslant P^{1/2}$, $(a,q) = 1$ and $|\beta| \leqslant 975(qP^3)^{-1}$, and write $\alpha = a/q + \beta$. Then according to Lemma 7.1, one has $|U_\varepsilon| \leqslant U$ for $\varepsilon \in \{0, 1\}$. Also, by Lemma 5.2 one has $|T_1| \leqslant |T_0|$. Thus we find that

$$\left| S_0(\alpha)^{7-t} S_1(\alpha)^t - T_0^{7-t} T_1^t \right| = \left| \sum_{\substack{i=0 \\ i+j \geqslant 1}}^{7-t} \sum_{j=0}^{t} \binom{7-t}{i} \binom{t}{j} T_0^{7-t-i} T_1^{t-j} U_0^i U_1^j \right|$$

$$\leqslant \sum_{\substack{i=0 \\ i+j \geqslant 1}}^{7-t} \sum_{j=0}^{t} \binom{7-t}{i} \binom{t}{j} |T_0|^{7-i-j} U^{i+j}$$

$$= (|T_0| + U)^7 - |T_0|^7 \leqslant 127(|T_0|^6 U + U^7). \tag{7.2}$$

Next, on writing

$$\Phi_{1,1} = \sum_{q \leqslant P^{1/2}} \sum_{\substack{a=1 \\ (a,q)=1}}^{q} \int_0^\infty |T_0|^6 U d\beta$$

and

$$\Phi_{1,2} = \frac{975}{P^3} \sum_{q \leqslant P^{1/2}} U^7,$$

we deduce from (7.1) and (7.2) that

$$|\Phi(n;t) - \Phi_1(n;t)| \leqslant 254(\Phi_{1,1} + \Phi_{1,2}). \tag{7.3}$$

CHAPTER 7. THE MAJOR ARC CONTRIBUTION

Recalling the definition (5.17) of $V(q)$, we see that

$$\Phi_{1,1} = 3 \times 10^6 \times 2^{-6} P^{1/2} \sum_{q \leqslant P^{1/2}} q^{1/4} V(q) \int_0^\infty |I(\beta)|^6 d\beta.$$

Thus, whenever $P \geqslant 10^{50}$ it follows from Lemma 5.5 that

$$\Phi_{1,1} < 3.03 \times 10^{10} P^{1/2} \log P \int_0^\infty |I(\beta)|^6 d\beta. \tag{7.4}$$

In order to evaluate the integral in (7.4), we observe that by making a change of variable in (4.1), we have

$$I(\beta) = 2P_0 J(16 P_0^4 \beta), \tag{7.5}$$

We therefore deduce from (4.3) that

$$|I(\beta)| \leqslant \min\{2P_0, (32\pi P_0^3 |\beta|)^{-1}\}, \tag{7.6}$$

whence we conclude that

$$\int_0^\infty |I(\beta)|^6 d\beta \leqslant \int_0^{(64\pi P_0^4)^{-1}} (2P_0)^6 d\beta + \int_{(64\pi P_0^4)^{-1}}^\infty (32\pi P_0^3 \beta)^{-6} d\beta$$
$$= \frac{P_0^2}{\pi} + \frac{P_0^2}{5\pi} = \frac{6P_0^2}{5\pi}.$$

On substituting the latter estimate into (7.4) and recalling (2.10), we find that whenever $P \geqslant 10^{50}$, one has

$$\Phi_{1,1} < 1.16 \times 10^{10} P^{5/2} \log P. \tag{7.7}$$

Meanwhile, one obtains with little effort the estimate

$$\Phi_{1,2} \leqslant 975 \times 3^7 \times 10^{42} P^{1/2} \sum_{q \leqslant P^{1/2}} q^{7/4}$$
$$< 2.2 \times 10^{48} P^{15/8}. \tag{7.8}$$

Consequently, on substituting (7.7) and (7.8) into (7.3), we conclude that whenever $P \geqslant 10^{50}$, one has

$$|\Phi(n;t) - \Phi_1(n;t)| \leqslant 254 P^3 (1.16 \times 10^{10} P^{-1/2} \log P + 2.2 \times 10^{48} P^{-9/8})$$
$$< 3.2 \times 10^{-6} P^3. \tag{7.9}$$

Our next step in the estimation of $\Phi(n;t)$ is to complete the integral in (7.1) to infinity. We therefore put

$$\Phi_2(n;t) = \sum_{q \leqslant P^{1/2}} \sum_{\substack{a=1 \\ (a,q)=1}}^q \int_{-\infty}^\infty T_0^{7-t} T_1^t e(-(a/q+\beta)n) d\beta, \tag{7.10}$$

and seek to bound $|\Phi_1(n;t) - \Phi_2(n;t)|$. By (7.6) and the trivial bound $|G_\varepsilon(q,a)| \leqslant q$, we obtain the upper bound

$$|\Phi_1(n;t) - \Phi_2(n;t)| \leqslant 2 \sum_{q \leqslant P^{1/2}} q \int_{975(qP^3)^{-1}}^{\infty} 2^{-7} |I(\beta)|^7 d\beta$$

$$\leqslant 2^{-6} \sum_{q \leqslant P^{1/2}} q \int_{P^{-7/2}}^{\infty} (P^3 \beta)^{-7} d\beta$$

$$\leqslant 2^{-6} (P^{1/2})^2 \times \frac{1}{6} < P.$$

Consequently, on substituting this estimate into (7.9), we deduce that

$$|\Phi(n;t) - \Phi_2(n;t)| < 3.3 \times 10^{-6} P^3. \tag{7.11}$$

We next observe that by (7.5) one has

$$\int_{-\infty}^{\infty} I(\beta)^7 e(-n\beta) d\beta = (2P_0)^7 \int_{-\infty}^{\infty} J(16P_0^4 \beta)^7 e(-n\beta) d\beta,$$

whence, on recalling (4.2) and making a change of variables, we deduce that

$$\int_{-\infty}^{\infty} I(\beta)^7 e(-n\beta) d\beta = 8P_0^3 K_7\big(n/(16P_0^4)\big).$$

On recalling the notation introduced in the statement of Lemma 6.2, we therefore deduce from (7.10) that

$$\Phi_2(n;t) = \mathfrak{S}(n, P^{1/2}; t) K_7\big(n/(16P_0^4)\big) P_0^3/16.$$

By hypothesis, we have $N - 4P^4 \leqslant n \leqslant N$, and thus it follows from (2.10) that $\nu - 1/4 \leqslant n/(16P_0^4) \leqslant \nu$. We therefore deduce from Corollary 4.3 that there exists a real number ν, with $23 < \nu < 64$, satisfying the property that $K_7(n/(16P_0^4)) \geqslant 0.01$. We fix this value of ν for the remainder of our argument. The truncated singular series may be bounded from below by reference to Lemma 6.2, and so we conclude that whenever $P \geqslant 10^{50}$, one has

$$\Phi_2(n;t) > (5.078 \times 0.01/16) P^3 > 0.003173 P^3 \quad \text{when } n \not\equiv 3 \pmod{5},$$

and

$$\Phi_2(n;t) > (1.269 \times 0.01/16) P^3 > 0.000793 P^3 \quad \text{when } n \equiv 3 \pmod{5}.$$

The proof of the lemma is therefore completed by combining the latter estimates with (7.11). □

We are now at last equipped to complete the proof of Lemma 2.1 by summing $\Phi(n;t)$ over a suitable set of integers n.

Proof of Lemma 2.1. — On recalling the definition (2.11) of $R(N; \mathfrak{M})$, it is evident that

$$R(N; \mathfrak{M}) = \sum_{m_1 \in \mathcal{M}_\eta(P^2)} \sum_{m_2 \in \mathcal{M}_0(3P^2/7)} \sum_{\substack{1 \leqslant w < P/6 \\ 2w+\zeta \notin \mathcal{W}(m_2)}} \Phi(\phi(N; m_1, m_2, w); t), \qquad (7.12)$$

where we write

$$\phi(N; m_1, m_2, w) = N - 2m_1^2 - 4m_2^2 - 24m_2(2w + \zeta)^2 - 6(2w + \zeta)^4. \qquad (7.13)$$

When m_1, m_2 and w satisfy the conditions imposed by the summations in (7.12), one has

$$N \geqslant \phi(N; m_1, m_2, w) \geqslant N - \left(2 + 4\left(\frac{3}{7}\right)^2 + 24\left(\frac{3}{7}\right)\left(\frac{1}{3}\right)^2 + 6\left(\frac{1}{3}\right)^4\right)P^4$$
$$> N - 4P^4.$$

Furthermore, in view of the conditions (2.3), (2.4) and (2.9), one has

$$\phi(N; m_1, m_2, w) \equiv t \pmod{16}.$$

It follows that for each choice of m_1, m_2 and w in the summations of (7.12), the integer $n = \phi(N; m_1, m_2, w)$ satisfies the hypotheses of Lemma 7.2, and thus we are able to employ the latter lemma to obtain a lower bound for $\Phi(\phi(N; m_1, m_2, w); t)$.

Observe next that for each fixed pair of integers, m_1 and m_2, there exists an integer $b = b(m_1, m_2)$ such that $\phi(N; m_1, m_2, w) \not\equiv 3 \pmod 5$ whenever $w \equiv b \pmod 5$. In order to verify this assertion, we note from (7.13) that the residue class of $\phi(N; m_1, m_2, w)$ modulo 5 may be shifted by an appropriate choice of w provided only that the polynomial $24 m_2 \xi^2 - 6\xi^4$ takes at least two distinct values modulo 5. But since for $5 \nmid \xi$ one has $\xi^2 \equiv \pm 1 \pmod 5$ and $\xi^4 \equiv 1 \pmod 5$, we find that the latter polynomial assumes the values 0, $\pm m_2 - 1$ modulo 5, whence our earlier assertion is immediate.

We now collect together the conclusions of the previous two paragraphs, deducing from Lemma 7.2 that there exists a real number ν with $23 < \nu < 64$, such that

$$R(N; \mathfrak{M}) > \sum_{m_1 \in \mathcal{M}_\eta(P^2)} \sum_{m_2 \in \mathcal{M}_0(3P^2/7)} (\Sigma_1 + \Sigma_2), \qquad (7.14)$$

where

$$\Sigma_1 = \sum_{\substack{1 \leqslant w < P/6 \\ 2w+\zeta \notin \mathcal{W}(m_2) \\ w \equiv b \pmod 5}} 0.00316 P^3 \quad \text{and} \quad \Sigma_2 = \sum_{\substack{1 \leqslant w < P/6 \\ 2w+\zeta \notin \mathcal{W}(m_2) \\ w \not\equiv b \pmod 5}} 0.000789 P^3.$$

On noting that by (2.5) one has $\operatorname{card}(\mathcal{W}(m_2)) \leqslant 3$, we find that

$$\Sigma_1 + \Sigma_2 \geqslant P^3 \left(0.00316\left(\frac{P}{30} - 4\right) + 0.000789\left(\frac{P}{6} - \frac{P}{30} - 5\right)\right).$$

Then on recalling the notation introduced in (2.13), it follows from (7.14) that for $P \geqslant 10^{50}$, one has
$$R(N; \mathfrak{M}) > 0.00021 M_\eta \widetilde{M_0} P^4,$$
and thus the proof of Lemma 2.1 is complete. □

CHAPTER 8

AN EXPLICIT VERSION OF WEYL'S INEQUALITY

In this section we describe the modifications to the argument of Deshouillers [7] required to establish the explicit version of Weyl's inequality recorded in Lemma 2.3 above. The idea of trading a small loss in the exponent of P in Weyl's bound for the benefit of a small constant, and smaller logarithmic factor, is due to Balasubramanian [2]. For the application at hand, the details of such an argument have been worked out in detail for $P \geqslant 10^{80}$ by Deshouillers [7]. We now modify the latter treatment so as to extend the validity of this estimate to the range $P \geqslant 10^{30}$, and also so as to avoid oppressive computations. We begin with an auxiliary lemma. Throughout this section, for the sake of concision we write $\beta = 0.036$ and $\gamma = 0.072$.

LEMMA 8.1. — *Define the multiplicative function f by*
$$f(n) = n^\beta \prod_{p \mid n} (1 - p^{-2\beta})^{1/2}.$$
Then for $P \geqslant 10^{30}$, one has
$$\sum_{n=1}^{\infty} \frac{1}{f(n)^2} \left(\sum_{\substack{l \mid n \\ l \leqslant P \\ n/l \leqslant P}} f(l) \right)^2 \leqslant 295900 P^{1-2\beta} \log P,$$
and
$$\sum_{1 \leqslant n \leqslant P} \frac{1}{f(n)} \leqslant 42.2 P^{1-\beta}.$$

Proof. — We follow the argument of the proof of Proposition 1 of Deshouillers [7], but now replace some infinite product evaluations by ones accessible to a hand-held calculator. We first observe that when $\sigma > 1$, one has
$$\zeta(\sigma) \leqslant \sum_{n=1}^{20} n^{-\sigma} + \int_{20}^{\infty} x^{-\sigma} dx = \sum_{n=1}^{20} n^{-\sigma} + \frac{20^{1-\sigma}}{\sigma - 1}.$$

Thus modest calculations reveal that
$$\zeta(1+\gamma) < 14.50, \quad \zeta(1+2\gamma) < 7.55, \quad \zeta(1+3\gamma) < 5.24,$$
$$\zeta(1+4\gamma) < 4.09, \quad \zeta(1+5\gamma) < 3.39.$$
Next we observe that for $0 < x < 3/4$, one has
$$\frac{1}{1-x} = 1 + x + x^2 + x^3 + \frac{x^4}{1-x} < 1 + x + x^2 + x^3 + 4x^4, \tag{8.1}$$
and furthermore, for $0 < x < 1$ one may derive the upper bound
$$\sqrt{1-x} < 1 - \frac{1}{2}x - \frac{1}{8}x^2 - \frac{1}{16}x^3 - \frac{5}{128}x^4 - \frac{7}{256}x^5, \tag{8.2}$$
whence for $0 < x < 3/4$ one obtains
$$\frac{1}{\sqrt{1-x}} < 1 + \frac{1}{2}x + \frac{3}{8}x^2 + \frac{5}{16}x^3 + \frac{35}{128}x^4 + \frac{63}{256}\frac{x^5}{1-x}$$
$$< 1 + \frac{1}{2}x + \frac{3}{8}x^2 + \frac{5}{16}x^3 + \frac{35}{128}x^4 + x^5. \tag{8.3}$$

It is useful at this point to recall some of the notation of the proof of Proposition 1 of Deshouillers [7]. We therefore define
$$G(p) = 1 + \frac{1}{p^{1+\gamma}(1-p^{-\gamma})}, \qquad \lambda(p) = \frac{1}{G(p)(1-p^{-\gamma})^{1/2}} - 1,$$
$$L(p) = 1 + \lambda(p)/p, \qquad V(p) = 1 + \frac{1}{p}\left(\frac{1}{(1-p^{-\gamma})^{1/2}} - 1\right).$$
We define also
$$K = \prod_p \left(1 + \frac{1}{p}\max\{\lambda(p)p^\gamma - 1, 0\}\right).$$
On observing that for $p \geqslant 59$, it follows from (8.1) that
$$G(p) < 1 + p^{-1-\gamma}(1 + p^{-\gamma} + p^{-2\gamma} + p^{-3\gamma} + 4p^{-4\gamma})$$
$$< (1 - p^{-1-5\gamma})^{-4} \prod_{j=1}^{4}(1 - p^{-1-j\gamma})^{-1},$$
we deduce from the previously calculated values of $\zeta(\sigma)$ that
$$\prod_p G(p) < \Big(\prod_{p \leqslant 53} G(p)\Big)\Big(\prod_{p \geqslant 59}(1-p^{-1-5\gamma})^{-4}\prod_{j=1}^{4}(1-p^{-1-j\gamma})^{-1}\Big)$$
$$= \Big(\prod_{p \leqslant 53} G(p)(1-p^{-1-5\gamma})^4 \prod_{j=1}^{4}(1-p^{-1-j\gamma})\Big)$$
$$\times \Big(\zeta(1+5\gamma)^4 \prod_{j=1}^{4}\zeta(1+j\gamma)\Big)$$
$$< 17456. \tag{8.4}$$

Similarly, one finds that for $p \geqslant 59$, as a consequence of (8.3) one has
$$V(p) < 1 + \frac{1}{2}p^{-1-\gamma} + \frac{3}{8}p^{-1-2\gamma} + \frac{5}{16}p^{-1-3\gamma} + \frac{35}{128}p^{-1-4\gamma} + p^{-1-5\gamma}$$
$$< \prod_{j=1}^{5}(1 - p^{-1-j\gamma})^{-e_j},$$
where we write $e_1 = 1/2$, $e_2 = 3/8$, $e_3 = 5/16$, $e_4 = 35/128$ and $e_5 = 1$. We therefore deduce that
$$\prod_{p} V(p) < \Big(\prod_{p \leqslant 53} V(p) \prod_{j=1}^{5}(1 - p^{-1-j\gamma})^{e_j}\Big)\Big(\prod_{j=1}^{5} \zeta(1 + j\gamma)^{e_j}\Big) < 40.6. \tag{8.5}$$
Since for each prime p one has $L(p) < V(p)$, it follows that
$$\prod_{p>3} L(p) < \Big(\prod_{5 \leqslant p \leqslant 53} L(p)\Big)\Big(\prod_{p \geqslant 59} V(p)\Big),$$
and thus one deduces similarly that
$$\prod_{p>3} L(p) < 3.95. \tag{8.6}$$

In order to evaluate K, we put $g(x) = ((1-x)^{-1/2} - 1)/x$, and observe that $g(x)$ is monotone increasing for $0 < x < 1$. Also one has $\lambda(p)p^\gamma < g(p^{-\gamma})$, and a modest calculation reveals that $g(800^{-\gamma}) < 1 < g(799^{-\gamma})$. Finally, we observe that the number of primes in the interval $[100, 200)$ is 21, and the number in $[200, 800)$ is 93. Drawing these observations together, we find that
$$K \leqslant \prod_{p<100}\Big(1 + \frac{1}{p}\max\{\lambda(p)p^\gamma - 1, 0\}\Big) \prod_{100<p<800}\Big(1 + \frac{1}{p}(g(p^{-\gamma}) - 1)\Big)$$
$$\leqslant \Big(1 + \frac{g(101^{-\gamma}) - 1}{101}\Big)^{21}\Big(1 + \frac{g(211^{-\gamma}) - 1}{211}\Big)^{93}$$
$$\times \prod_{p \leqslant 97}\Big(1 + \frac{1}{p}\max\{\lambda(p)p^\gamma - 1, 0\}\Big)$$
$$\leqslant 1.111 \prod_{17 \leqslant p \leqslant 97}\Big(1 + \frac{1}{p}(\lambda(p)p^\gamma - 1)\Big) < 1.19. \tag{8.7}$$

Having eliminated the burdensome computations of [7] with the above discussion, we now complete the proof of the lemma along the same lines as the argument of the proof of Proposition 1 of Deshouillers [7]. We begin by observing that the upper bound for S provided in the second display of p.296 of [7] yields
$$\sum_{n=1}^{\infty} \frac{1}{f(n)^2}\Big(\sum_{\substack{l \mid n \\ l \leqslant P \\ n/l \leqslant P}} f(l)\Big)^2 \leqslant \frac{2P^{1-\gamma}}{1-\gamma}\Big(\prod_p G(p)\Big)T, \tag{8.8}$$

where T is an expression bounded in the third display of p.297 of [7] in the form

$$T \leqslant \frac{1}{G(2)} U(2,1,1) + \Lambda(2) U(2,2,1) + \Lambda(2) U(2,1,2)$$
$$+ U(1,1,1) + \Lambda(2) U(1,2,1) + \Lambda(2) U(1,1,2). \quad (8.9)$$

In this latter expression, the number $\Lambda(2)$ is given by

$$\Lambda(2) = (1 - 2^{-\gamma})^{-1/2}/G(2), \quad (8.10)$$

and when $\varepsilon, \eta_1, \eta_2 \in \{1, 2\}$, the argument on p.297 of [7] establishes the bound

$$U(\varepsilon, \eta_1, \eta_2) \leqslant \sum_{\substack{1 \leqslant r \leqslant P \\ r \equiv \varepsilon \pmod{2}}} \bigl(W_1(r) + W_2(r) + W_3(r) + W_4(r)\bigr), \quad (8.11)$$

in which

$$W_1(r) = \frac{P}{4(1-\beta)r} \prod_{p>3} L(p)^2, \quad (8.12)$$

$$W_2(r) = \frac{K}{2(1-\beta)(1-2\beta)} \Bigl(\prod_{p>3} L(p)\Bigr) \Bigl(\frac{P}{r}\Bigr)^{1-2\beta}, \quad (8.13)$$

$$W_3(r) = \frac{K}{2(1-3\beta)(1-2\beta)} \Bigl(\prod_{p>3} L(p)\Bigr) \Bigl(\frac{P}{r}\Bigr)^{1-2\beta}, \quad (8.14)$$

$$W_4(r) = \frac{K^2}{(1-3\beta)(1-4\beta)} \Bigl(\frac{P}{r}\Bigr)^{1-4\beta}. \quad (8.15)$$

To these estimates we add the bounds

$$\sum_{\substack{1 \leqslant r \leqslant P \\ r \equiv \varepsilon \pmod{2}}} \frac{1}{r} \leqslant \Bigl(1 + \frac{1}{2} - \frac{\log 2}{2}\Bigr) + \frac{1}{2} \log P, \quad (8.16)$$

and

$$\sum_{\substack{1 \leqslant r \leqslant P \\ r \equiv \varepsilon \pmod{2}}} r^{-\delta} \leqslant 1 + \frac{P^{1-\delta}}{2(1-\delta)} \leqslant 7 P^{1-\delta}, \quad (8.17)$$

valid for $0 < \delta \leqslant 1 - 2\beta$ and $P \geqslant 10^{30}$, which follow from the discussion at the top of p.298 of [7].

On substituting (8.16) and (8.17) into (8.12)-(8.15), we deduce from (8.6) and (8.7) that whenever $P \geqslant 10^{30}$, one has

$$\sum_{\substack{1 \leqslant r \leqslant P \\ r \equiv \varepsilon \pmod{2}}} W_i(r) < C_i P \log P \quad (1 \leqslant i \leqslant 4),$$

where $C_1 = 2.0908$, $C_2 = 0.2663$, $C_3 = 0.2878$, $C_4 = 0.1880$. Thus one deduces from (8.11) the estimate

$$U(\varepsilon, \eta_1, \eta_2) \leqslant 2.833 P \log P.$$

On substituting this estimate into (8.9) together with (8.10), we therefore deduce that
$$T \leqslant 7.865 P \log P,$$
whence by (8.4) and (8.8),
$$\sum_{n=1}^{\infty} \frac{1}{f(n)^2} \left(\sum_{\substack{l \mid n \\ l \leqslant P \\ n/l \leqslant P}} f(l) \right)^2 \leqslant 295900 P \log P.$$

This completes the proof of the first assertion of the lemma.

In order to establish the final assertion of the lemma, we apply Lemma 1 of [**7**] with $a = \beta$, $b = 0$, $h = k = 1$ and $\rho(p) = (1 - p^{-\gamma})^{-1/2} - 1$. Thus we deduce that
$$\sum_{1 \leqslant n \leqslant P} \frac{1}{f(n)} \leqslant \frac{P^{1-\beta}}{1-\beta} \prod_p V(p),$$
whence the desired conclusion follows immediately from (8.5).

We now establish Lemma 2.3. Suppose that $\alpha \in \mathfrak{m}$. By Dirichlet's theorem on diophantine approximation, we can find $a \in \mathbb{Z}$ and $q \in \mathbb{N}$ with $(a, q) = 1$, $1 \leqslant q \leqslant P^3/975$ and $|q\alpha - a| \leqslant 975 P^{-3}$. Then it follows from the definition of $\mathfrak{m}$ that necessarily $q > P^{1/2}$. We divide our argument into two cases according to the relative sizes of q and P. In order to facilitate this discussion we define the parameter $\tau(P)$ by
$$\tau(P) = \begin{cases} 1/2, & \text{when } 10^{30} \leqslant P < 10^{53}, \\ 2 \times 10^6, & \text{when } P \geqslant 10^{53}. \end{cases}$$

Suppose first that $P^{1/2} < q \leqslant \tau(P)P$. We apply the argument of the proof of Proposition 5.1 of Deshouillers and Dress [**9**], but now extend the range of validity down to $P \geqslant 10^{30}$. Recalling the notation introduced at the start of §7, we find from equation (2.4.2) of [**9**] that for $\varepsilon \in \{0, 1\}$, one has
$$|S_\varepsilon(\alpha) - (2q)^{-1} G_\varepsilon(q, a) I(\beta)| \leqslant 2.7 \times 10^6 q^{1/4} P^{1/2} + 61 q^{3/4}(\log q + 1), \qquad (8.18)$$
valid for any positive number P. By combining the bound (5.7) with the conclusion of Lemma 5.2, we obtain the estimate
$$|G_\varepsilon(q, a)| \leqslant 9 q^{3/4}. \qquad (8.19)$$
Thus, on making use of the trivial bound $|I(\beta)| \leqslant 2P_0 \leqslant 2P + 2$, we conclude from (8.16) and (8.17) that when $P \geqslant 10^{30}$ and $P^{1/2} < q \leqslant \tau(P)P$, one has
$$|S_\varepsilon(\alpha)| \leqslant 9(P+1)q^{-1/4} + 2.7 \times 10^6 q^{1/4} P^{1/2} + 61 q^{3/4}(\log q + 1)$$
$$\leqslant \kappa_1(P) P^{0.884} (\log P)^{0.25}, \qquad (8.20)$$
where
$$\kappa_1(P) = \begin{cases} 77, & \text{when } 10^{30} \leqslant P < 10^{53}, \\ 14, & \text{when } P \geqslant 10^{53}. \end{cases}$$

Suppose next that $\tau(P)P < q \leqslant P^3/975$. Here we note first that the argument of the proof of Proposition 2 of [**7**] yields the estimate
$$\sum_{1 \leqslant n \leqslant P^3/27} \min\Big\{P, \frac{1}{2\|384n\alpha\|}\Big\} \leqslant \Big(\frac{P^3}{27q'} + 1\Big)\big(28000P + q'\log(2q')\big),$$
where we write $q' = q/(384, q)$. On noting that the maximum of the latter expression for $q \in [\tau(P)P, P^3/975]$ is achieved at one of the end points of this interval, we deduce that
$$\sum_{1 \leqslant n \leqslant P^3/27} \min\Big\{P, \frac{1}{2\|384n\alpha\|}\Big\} \leqslant A(P)P^3 \log P, \tag{8.21}$$
where
$$A(P) = \begin{cases} 11530, & \text{when } 10^{30} \leqslant P < 10^{53}, \\ 0.1123, & \text{when } P \geqslant 10^{53}. \end{cases}$$

Finally, we work through the argument of §3 of [**7**], though now replacing use of Propositions 1 and 2 in the latter by the conclusion of Lemma 8.1 above, and (8.21), respectively. In this way one bounds an exponential sum essentially equal to $S_\varepsilon(\alpha)$ in terms of auxiliary sums $T_i(\alpha)$ ($1 \leqslant i \leqslant 5$), the definitions of which we suppress in the interest of saving space.

First one finds that
$$T_5(\alpha) \leqslant \Big(\frac{P^3}{27}\Big)^{0.072} \sum_{1 \leqslant n \leqslant P^3/27} \min\Big\{P, \frac{1}{2\|384n\alpha\|}\Big\}$$
$$\leqslant 0.7888 A(P) P^{3.216} \log P.$$

Next,
$$T_4(\alpha) \leqslant P \sum_{1 \leqslant n \leqslant P^2/4} f(n)^2 + 2T_5(\alpha)$$
$$\leqslant P(P^2/4)^{1.072} + 1.5776 A(P) P^{3.216} \log P,$$
whence for $P \geqslant 10^{30}$ one has
$$T_4(\alpha) \leqslant 1.5781 A(P) P^{3.216} \log P.$$

At the next stage we obtain
$$|T_3(\alpha)|^2 \leqslant \Big(\sum_{n=1}^{\infty} \frac{1}{f(n)^2} \Big(\sum_{\substack{l \mid n \\ l \leqslant P \\ n/l \leqslant P}} f(l)\Big)^2\Big) T_4(\alpha),$$
whence
$$|T_3(\alpha)| \leqslant \big(295900 P^{1.928} \log P\big)^{1/2} \big(1.5781 A(P) P^{3.216} \log P\big)^{1/2}$$
$$\leqslant 683.4 A(P)^{1/2} P^{2.572} \log P.$$

Then we arrive at the estimate

$$T_2(\alpha) \leqslant P \sum_{h_1=1}^{P} f(h_1) + 2T_3(\alpha)$$

$$\leqslant P(P^{1.036}) + 1366.8 A(P)^{1/2} P^{2.572} \log P,$$

so that for $P \geqslant 10^{30}$ one obtains

$$T_2(\alpha) \leqslant 1367 A(P)^{1/2} P^{2.572} \log P.$$

Next we have

$$|T_1(\alpha)|^2 \leqslant \Big(\sum_{h_1=1}^{P} \frac{1}{f(n)} \Big) T_2(\alpha)$$

$$\leqslant (42.2 P^{0.964})(1367 A(P)^{1/2} P^{2.572} \log P),$$

whence

$$|T_1(\alpha)| \leqslant 240.2 A(P)^{1/4} P^{1.768} (\log P)^{1/2}.$$

Finally, we have

$$\Big| \sum_{x=P+1}^{2P} e(\alpha(2x+\varepsilon)^4) \Big|^2 \leqslant P + 2T_1(\alpha),$$

whence for $P \geqslant 10^{30}$ we obtain

$$\Big| \sum_{x=P+1}^{2P} e(\alpha(2x+\varepsilon)^4) \Big| \leqslant 21.92 A(P)^{1/8} P^{0.884} (\log P)^{1/4}.$$

On recalling the definition of $A(P)$, and accounting for the endpoints in the summation of (2.1) by a trivial estimate, we deduce that in this second case in which $\tau(P)P < q \leqslant P^3/975$, one has

$$|S_\varepsilon(\alpha)| \leqslant \kappa_2(P) P^{0.884} (\log P)^{0.25},$$

where

$$\kappa_2(P) = \begin{cases} 70.6, & \text{when } 10^{30} \leqslant P < 10^{53}, \\ 16.68, & \text{when } P \geqslant 10^{53}. \end{cases}$$

The conclusion of Lemma 2.3 therefore follows on combining the latter estimate with (8.20) above.

CHAPTER 9

AN AUXILIARY BOUND FOR THE DIVISOR FUNCTION

The final ingredients in our proof of Theorem 2 are the mean value estimates recorded in Lemmata 2.4 and 2.5, and indeed it is these estimates that provide the crucial additional power required to establish a conclusion with only sixteen biquadrates. In order to transform the ideas of Kawada and Wooley [15] into technology appropriate to the present application, one must provide strong explicit estimates for certain averages of the divisor function evaluated on biquadratic polynomials. Here we follow the trail laid down by Deshouillers and Dress [8], and developed by Landreau in his thesis [16]. Write $\tau(n)$ for the number of divisors of n. Then we seek an upper bound for $\tau(n)$ of the shape

$$\tau(n) \leqslant C \sum_{\substack{d \mid n \\ d \leqslant n^{1/4}}} g(d), \tag{9.1}$$

for some appropriate multiplicative function g and constant C. In discussing this topic, we must draw a balance between the objectives of keeping our exposition reasonably short, and at the same time achieving a relatively strong bound for the function g. Motivated by such considerations, we define the multiplicative function g for prime powers p^l by taking

$$g(p^l) = \begin{cases} 3, & \text{when } l = 1, 2, \\ 27, & \text{when } l = 7, 9, 11, \\ 9, & \text{otherwise.} \end{cases} \tag{9.2}$$

With the function g defined in this way, we establish the validity of the upper bound (9.1) with $C = 8$. We remark that the latter value of C is best possible, for when n is the product of three distinct primes of almost the same size, the inequality (9.1) evidently cannot hold with $C < 8$. However, with a little further effort, one can show that, with the exception of such special integers n, the upper bound (9.1) is indeed valid with a suitable constant $C < 8$.

In order to facilitate our analysis, we introduce some slightly unconventional vector notation. We note, in particular, that in any vector that appears in our discussion below, we implicitly assume that every component is a non-negative integer. When k is a natural number, we define the integer $\boldsymbol{x}^{\boldsymbol{a}}$ corresponding to the vectors $\boldsymbol{x} = (x_1, \ldots, x_k)$ and $\boldsymbol{a} = (a_1, \ldots, a_k)$ by

$$\boldsymbol{x}^{\boldsymbol{a}} = \prod_{j=1}^{k} x_j^{a_j}. \tag{9.3}$$

When $\boldsymbol{b} = (b_1, \ldots, b_k)$, we define also

$$\widetilde{\tau}(\boldsymbol{b}) = \prod_{j=1}^{k}(b_j + 1),$$

and, when p is a prime number, we define also

$$\widetilde{g}(\boldsymbol{b}) = \prod_{j=1}^{k} g(p^{b_j}).$$

We use the notation $\boldsymbol{b} \leqslant \boldsymbol{a}$ as shorthand for the condition that $\boldsymbol{a} = (a_1, \ldots, a_k)$, $\boldsymbol{b} = (b_1, \ldots, b_k)$ and $0 \leqslant b_j \leqslant a_j$ for $1 \leqslant j \leqslant k$, and then define

$$S(\boldsymbol{a}; \boldsymbol{x}) = \{\boldsymbol{b} \leqslant \boldsymbol{a} \; ; \; \boldsymbol{x}^{4\boldsymbol{b}} \leqslant \boldsymbol{x}^{\boldsymbol{a}}\}, \qquad G(\boldsymbol{a}; \boldsymbol{x}) = \sum_{\boldsymbol{b} \in S(\boldsymbol{a};\boldsymbol{x})} \widetilde{g}(\boldsymbol{b}),$$

$$C(\boldsymbol{a}; \boldsymbol{x}) = \widetilde{\tau}(\boldsymbol{a})/G(\boldsymbol{a}; \boldsymbol{x}), \qquad C(\boldsymbol{a}) = \sup_{\boldsymbol{x} \in \mathbb{N}^k} C(\boldsymbol{a}; \boldsymbol{x}).$$

Finally, it is convenient to write $V = \bigcup_{k \in \mathbb{N}} \mathbb{N}^k$ and $V_1 = \bigcup_{k \in \mathbb{N}} \{1, 2, 4, 6, 10\}^k$.

Some preliminary observations are in order concerning the notation recorded in the previous paragraph. First, since the zero vector is plainly contained in $S(\boldsymbol{a}; \boldsymbol{x})$, one finds that for every choice of $\boldsymbol{a}$ and $\boldsymbol{x}$ one has $G(\boldsymbol{a}; \boldsymbol{x}) \geqslant 1$, whence

$$C(\boldsymbol{a}) \leqslant \widetilde{\tau}(\boldsymbol{a}). \tag{9.4}$$

We observe also that the assertion (9.1) may be translated into a related statement concerning the function $C(\boldsymbol{a})$. Indeed, it is this observation that motivates our choice of notation. When $n = 1$, of course, the inequality (9.1) is trivial whenever $C \geqslant 1$, and thus the desired relation is not at issue. Suppose then that $n > 1$, and let $n = \prod_{j=1}^{k} p_j^{a_j}$ be the canonical prime factorisation of n. We put $\boldsymbol{a} = (a_1, \ldots, a_k)$ and $\boldsymbol{p} = (p_1, \ldots, p_k)$, and note that $\tau(n) = \widetilde{\tau}(\boldsymbol{a})$ and

$$\sum_{\substack{d \mid n \\ d \leqslant n^{1/4}}} g(d) = G(\boldsymbol{a}; \boldsymbol{p}).$$

Consequently, the inequality (9.1) is valid with

$$C = \sup_{\boldsymbol{a} \in V} C(\boldsymbol{a}), \tag{9.5}$$

MÉMOIRES DE LA SMF 100

provided that this supremum exists. As we have already pointed out implicitly in our opening remarks, one has $C(\boldsymbol{a}) = 8$ when $\boldsymbol{a} = (1,1,1)$. In order to establish (9.1) with $C = 8$, therefore, it suffices to show that for all $\boldsymbol{a} \in V$, one has $C(\boldsymbol{a}) \leqslant 8$.

As our first step towards the goal announced in the previous paragraph, we show that
$$\sup_{\boldsymbol{a} \in V} C(\boldsymbol{a}) = \sup_{\boldsymbol{b} \in V_1} C(\boldsymbol{b}). \tag{9.6}$$
This simplification is achieved as a consequence of Lemmata 9.1, 9.2 and 9.3 below. It is useful at this point to introduce a convention that simplifies our discussion of the vectors which play a central role in our argument. Hereafter, when $\boldsymbol{a} = (a_1, \ldots, a_k)$ and $\boldsymbol{a}' = (a'_1, \ldots, a'_{k'})$, we abbreviate the vector $(a_1, \ldots, a_k, a'_1, \ldots, a'_{k'})$ to $(\boldsymbol{a}, \boldsymbol{a}')$. Furthermore, where confusion is easily avoided, when f is a function of a vector variable $\boldsymbol{x}$, we abbreviate $f((\boldsymbol{a}, \boldsymbol{a}'))$ simply to $f(\boldsymbol{a}, \boldsymbol{a}')$. We extend such conventions to write $(\boldsymbol{a}, a)$ in place of $(a_1, \ldots, a_k, a)$, and $f(\boldsymbol{a}, a)$ in place of $f((\boldsymbol{a}, a))$. Also, on occasion, we write a for (a), and $f(a)$ for $f((a))$.

LEMMA 9.1. — *For any $\boldsymbol{a}_1, \boldsymbol{a}_2 \in V$, one has*
$$C(\boldsymbol{a}_1, \boldsymbol{a}_2) \leqslant C(\boldsymbol{a}_1) C(\boldsymbol{a}_2).$$

Proof. — Suppose that $\boldsymbol{a}_i \in \mathbb{N}^{k_i}$ for $i = 1, 2$. Then in view of (9.3), whenever $\boldsymbol{x}_i \in \mathbb{N}^{k_i}$ ($i = 1, 2$), one has
$$(\boldsymbol{x}_1, \boldsymbol{x}_2)^{(\boldsymbol{a}_1, \boldsymbol{a}_2)} = \boldsymbol{x}_1^{\boldsymbol{a}_1} \boldsymbol{x}_2^{\boldsymbol{a}_2}.$$
But it is apparent from our definitions that whenever $\boldsymbol{b}_i \in S(\boldsymbol{a}_i; \boldsymbol{x}_i)$ ($i = 1, 2$), then one has
$$(\boldsymbol{b}_1, \boldsymbol{b}_2) \in S((\boldsymbol{a}_1, \boldsymbol{a}_2); (\boldsymbol{x}_1, \boldsymbol{x}_2)),$$
and hence
$$G((\boldsymbol{a}_1, \boldsymbol{a}_2); (\boldsymbol{x}_1, \boldsymbol{x}_2)) \geqslant \sum_{\boldsymbol{b}_1 \in S(\boldsymbol{a}_1; \boldsymbol{x}_1)} \sum_{\boldsymbol{b}_2 \in S(\boldsymbol{a}_2; \boldsymbol{x}_2)} \widetilde{g}(\boldsymbol{b}_1, \boldsymbol{b}_2).$$
But
$$\widetilde{g}(\boldsymbol{b}_1, \boldsymbol{b}_2) = \widetilde{g}(\boldsymbol{b}_1) \widetilde{g}(\boldsymbol{b}_2) \quad \text{and} \quad \widetilde{\tau}(\boldsymbol{a}_1, \boldsymbol{a}_2) = \widetilde{\tau}(\boldsymbol{a}_1) \widetilde{\tau}(\boldsymbol{a}_2), \tag{9.7}$$
and thus we deduce that
$$G((\boldsymbol{a}_1, \boldsymbol{a}_2); (\boldsymbol{x}_1, \boldsymbol{x}_2)) \geqslant G(\boldsymbol{a}_1; \boldsymbol{x}_1) G(\boldsymbol{a}_2; \boldsymbol{x}_2),$$
and
$$C((\boldsymbol{a}_1, \boldsymbol{a}_2); (\boldsymbol{x}_1, \boldsymbol{x}_2)) \leqslant C(\boldsymbol{a}_1; \boldsymbol{x}_1) C(\boldsymbol{a}_2; \boldsymbol{x}_2).$$
The conclusion of the lemma is immediate from the latter inequality. □

LEMMA 9.2. — *Whenever a is an integer with $a \geqslant 12$, one has $C(a) \leqslant 1$.*

Proof. — Suppose that a is an integer with $a \geqslant 12$, and write $a = 4m + r$ with $0 \leqslant r \leqslant 3$. In particular, one has $m \geqslant 3$. But plainly, for every natural number x one has that $(0), (1), \ldots, (m) \in S(a;x)$, and further, when $j \geqslant 3$ one has $\widetilde{g}(j) \geqslant 9$. Thus we deduce that

$$G(a;x) \geqslant \sum_{j=0}^{m} \widetilde{g}(j) \geqslant 1 + 3 + 3 + 9(m-2)$$
$$\geqslant 4m + 4 \geqslant a + 1 = \widetilde{\tau}(a).$$

We therefore conclude that for every natural number x, one has $C(a;x) \leqslant 1$, and the conclusion of the lemma follows immediately. □

Before announcing the next lemma, we extend our conventions concerning vector notation by introducing the *empty vector* $\mathfrak{z}$, by which we mean the vector having no components. Plainly, for every vector $\boldsymbol{a}$ one has $(\boldsymbol{a}, \mathfrak{z}) = (\mathfrak{z}, \boldsymbol{a}) = \boldsymbol{a}$.

LEMMA 9.3. — *Suppose that $\boldsymbol{a}$ is a vector in $V \cup \{\mathfrak{z}\}$. Then the following inequalities hold:*

$$C(\boldsymbol{a}, 3) \leqslant C(\boldsymbol{a}, 1, 1), \qquad C(\boldsymbol{a}, 5) \leqslant C(\boldsymbol{a}, 1, 2),$$
$$C(\boldsymbol{a}, 7) \leqslant C(\boldsymbol{a}, 1, 1, 1), \qquad C(\boldsymbol{a}, 8) \leqslant C(\boldsymbol{a}, 2, 2),$$
$$C(\boldsymbol{a}, 9) \leqslant C(\boldsymbol{a}, 1, 4), \qquad C(\boldsymbol{a}, 11) \leqslant C(\boldsymbol{a}, 1, 1, 2).$$

Proof. — For ease of exposition in the discussion that follows, we restrict our proof to vectors $\boldsymbol{a}$ in V. However, an inspection of our argument will reveal that it applies equally well when $\boldsymbol{a}$ is empty. Suppose then that $k \in \mathbb{N}$ and $\boldsymbol{a} \in \mathbb{N}^k$, and consider arbitrary vectors $\boldsymbol{x} \in \mathbb{N}^k$ and $x \in \mathbb{N}$.

We begin by examining the simplest case, which concerns the inequality $C(\boldsymbol{a}, 3) \leqslant C(\boldsymbol{a}, 1, 1)$, since this sets the scene for the remaining cases. Define the map $\phi_3 : \{0, 1, 2, 3\} \to \{0, 1\}^2$ by taking

$$\phi_3(0) = (0,0), \quad \phi_3(1) = (1,0), \quad \phi_3(2) = (0,1), \quad \phi_3(3) = (1,1).$$

On recalling our notation (9.3), we see that whenever $b \in \{0, 1, 2, 3\}$, one has

$$x^b = (x, x^2)^{\phi_3(b)}. \tag{9.8}$$

Next, we define the map $\widetilde{\phi}_3$ for $(\boldsymbol{b}, b) \in S((\boldsymbol{a}, 3); (\boldsymbol{x}, x))$ by taking

$$\widetilde{\phi}_3(\boldsymbol{b}, b) = (\boldsymbol{b}, \phi_3(b)).$$

In view of the relation (9.8), and the observation that ϕ_3 plainly provides a bijection from $\{0, 1, 2, 3\}$ to $\{0, 1\}^2$, one readily confirms that $\widetilde{\phi}_3$ is a bijection from $S((\boldsymbol{a}, 3); (\boldsymbol{x}, x))$ to $S((\boldsymbol{a}, 1, 1); (\boldsymbol{x}, x, x^2))$. Furthermore, on recalling (9.2), one finds that for each $b \in \{0, 1, 2, 3\}$, one has $\widetilde{g}(b) = \widetilde{g}(\phi_3(b))$, whence also $\widetilde{g}(\boldsymbol{b}, b) = \widetilde{g}(\widetilde{\phi}_3(\boldsymbol{b}, b))$ for $(\boldsymbol{b}, b) \in S((\boldsymbol{a}, 3); (\boldsymbol{x}, x))$. We therefore deduce that

$$G((\boldsymbol{a}, 3); (\boldsymbol{x}, x)) = \sum_{(\boldsymbol{b},b) \in S((\boldsymbol{a},3);(\boldsymbol{x},x))} \widetilde{g}(\widetilde{\phi}_3(\boldsymbol{b}, b))$$

$$= \sum_{\boldsymbol{c} \in S((\boldsymbol{a},1,1);(\boldsymbol{x},x,x^2))} \widetilde{g}(\boldsymbol{c})$$

$$= G((\boldsymbol{a}, 1, 1); (\boldsymbol{x}, x, x^2)).$$

But the definition of $\widetilde{\tau}$ reveals that $\widetilde{\tau}(\boldsymbol{a}, 3) = 4\widetilde{\tau}(\boldsymbol{a}) = \widetilde{\tau}(\boldsymbol{a}, 1, 1)$, and thus it follows that

$$C((\boldsymbol{a}, 3); (\boldsymbol{x}, x)) = C((\boldsymbol{a}, 1, 1); (\boldsymbol{x}, x, x^2)).$$

We therefore conclude that

$$C(\boldsymbol{a}, 3) = \sup_{(\boldsymbol{x},x) \in \mathbb{N}^{k+1}} C((\boldsymbol{a}, 1, 1); (\boldsymbol{x}, x, x^2))$$

$$\leqslant \sup_{\boldsymbol{y} \in \mathbb{N}^{k+2}} C((\boldsymbol{a}, 1, 1); \boldsymbol{y}) = C(\boldsymbol{a}, 1, 1),$$

and this establishes the desired inequality in the first case.

We establish the remaining inequalities recorded in the statement of the lemma by a similar argument. We begin by defining analogues of the map ϕ_3 appropriate to each case. For $a = 5, 7, 8, 9$, define the map ϕ_a by means of the entries in the following tables.

b	0	1	2	3	4	5
$\phi_5(b)$	$(0,0)$	$(1,0)$	$(0,1)$	$(1,1)$	$(0,2)$	$(1,2)$

b	0	1	2	3	4	5	6	7
$\phi_7(b)$	$(0,0,0)$	$(1,0,0)$	$(0,1,0)$	$(1,1,0)$	$(0,0,1)$	$(1,0,1)$	$(0,1,1)$	$(1,1,1)$

b	0	1	2	3	4	5	6	7	8
$\phi_8(b)$	$(0,0)$	$(1,0)$	$(2,0)$	$(0,1)$	$(1,1)$	$(2,1)$	$(0,2)$	$(1,2)$	$(2,2)$

b	0	1	2	3	4	5	6	7	8	9
$\phi_9(b)$	$(0,0)$	$(1,0)$	$(0,1)$	$(1,1)$	$(0,2)$	$(1,2)$	$(0,3)$	$(1,3)$	$(0,4)$	$(1,4)$

Also, define the map ϕ_{11} by taking

$$\phi_{11}(b) = \begin{cases} \phi_7(b), & \text{for } 0 \leqslant b \leqslant 7, \\ \phi_7(b-4) + (0,0,1), & \text{for } 8 \leqslant b \leqslant 11. \end{cases}$$

Next put

$$\boldsymbol{x}_5 = \boldsymbol{x}_9 = (x, x^2), \quad \boldsymbol{x}_7 = \boldsymbol{x}_{11} = (x, x^2, x^4), \quad \boldsymbol{x}_8 = (x, x^3).$$

Then one may verify that whenever $a \in \{5, 7, 8, 9, 11\}$ and $0 \leqslant b \leqslant a$, then one has

$$x^b = \boldsymbol{x}_a^{\phi_a(b)} \quad \text{and} \quad \widetilde{g}(b) \geqslant \widetilde{g}(\phi_a(b)).$$

For the sake of brevity, we now write $\boldsymbol{a}_a = \phi_a(\boldsymbol{a})$ and
$$S_a = S((\boldsymbol{a}, a); (\boldsymbol{x}, x)), \qquad S'_a = S((\boldsymbol{a}, \boldsymbol{a}_a); (\boldsymbol{x}, \boldsymbol{x}_a)),$$
and we define the map $\widetilde{\phi}_a$ for $(\boldsymbol{b}, b) \in S_a$ by taking
$$\widetilde{\phi}_a(\boldsymbol{b}, b) = (\boldsymbol{b}, \phi_a(b)).$$
Then one may confirm that for each $a \in \{5, 7, 8, 9, 11\}$, the map $\widetilde{\phi}_a$ provides a bijection from S_a to S'_a. Thus we deduce that
$$G((\boldsymbol{a}, a); (\boldsymbol{x}, x)) \geqslant \sum_{(\boldsymbol{b}, b) \in S_a} \widetilde{g}(\widetilde{\phi}_a(\boldsymbol{b}, b)) = \sum_{\boldsymbol{b}' \in S'_a} \widetilde{g}(\boldsymbol{b}')$$
$$= G((\boldsymbol{a}, \boldsymbol{a}_a); (\boldsymbol{x}, \boldsymbol{x}_a)).$$
But $\widetilde{\tau}(\boldsymbol{a}, a) = \widetilde{\tau}(\boldsymbol{a}, \boldsymbol{a}_a)$ for $a \in \{5, 7, 8, 9, 11\}$, and so it follows that
$$C((\boldsymbol{a}, a); (\boldsymbol{x}, x)) \leqslant C((\boldsymbol{a}, \boldsymbol{a}_a); (\boldsymbol{x}, \boldsymbol{x}_a)).$$
Consequently, just as in the first case considered in this proof, one deduces that $C(\boldsymbol{a}, a) \leqslant C(\boldsymbol{a}, \boldsymbol{a}_a)$ for $a \in \{5, 7, 8, 9, 11\}$. This completes the proof of the lemma. □

We are now equipped to confirm the relation (9.6). Observe first that the value of $C(\boldsymbol{a})$ is independent of the order of the components of $\boldsymbol{a}$. If all of the components of $\boldsymbol{a}$ exceed 11, then it follows from Lemmata 9.1 and 9.2 that $C(\boldsymbol{a}) \leqslant 1 \leqslant C(1)$, whence such vectors play no role in determining the supremum on the left hand side of (9.6). Next, again by Lemmata 9.1 and 9.2, one may delete any exponent exceeding 11 from $\boldsymbol{a}$ without increasing the value of $C(\boldsymbol{a})$. By Lemmata 9.1 and 9.3, moreover, any component of $\boldsymbol{a}$ lying in the set $\{3, 5, 7, 8, 9, 11\}$ may be replaced by a vector from the set $\{(1, 1), (1, 2), (1, 1, 1), (2, 2), (1, 4), (1, 1, 2)\}$, again without increasing the value of $C(\boldsymbol{a})$. We therefore conclude that whenever $\boldsymbol{a} \in V$, there exists a vector $\boldsymbol{b} \in V_1$ with the property that $C(\boldsymbol{a}) \leqslant C(\boldsymbol{b})$, whence the desired conclusion (9.6) follows immediately.

Our second step towards the evaluation of the supremum defined in (9.5) involves removing from consideration all vectors $\boldsymbol{a}$ with the property that none of their components are equal to 1. To this end, we introduce the sets
$$V_2 = \bigcup_{k \in \mathbb{N}} \{2, 4, 6, 10\}^k \qquad \text{and} \qquad V'_2 = V_2 \cup \{\mathfrak{z}\}.$$
Also, when $\boldsymbol{a} \in V_2$, we define the function $f(\boldsymbol{a})$ by
$$f(\boldsymbol{a}) = \frac{1}{2} \sum_{\boldsymbol{b} \leqslant \boldsymbol{a}/2} \min\{\widetilde{g}(\boldsymbol{b}), \widetilde{g}(\boldsymbol{a}/2 - \boldsymbol{b})\},$$
and we define the function $F(\boldsymbol{a})$ for $\boldsymbol{a} \in V'_2$ by
$$F(\boldsymbol{a}) = \begin{cases} \widetilde{\tau}(\boldsymbol{a})/f(\boldsymbol{a}), & \text{when } \boldsymbol{a} \in V_2, \\ 2, & \text{when } \boldsymbol{a} = \mathfrak{z}. \end{cases}$$

The next lemma presents the relation between $F(\boldsymbol{a})$ and $C(\boldsymbol{a})$ crucial to the success of this next phase of our argument.

LEMMA 9.4. — *Whenever $\boldsymbol{a} \in V_2$, one has $C(\boldsymbol{a}) \leqslant F(\boldsymbol{a})$.*

Proof. — Suppose that $\boldsymbol{a}$ is a k-dimensional vector in V_2. Then for any $\boldsymbol{x} \in \mathbb{N}^k$, and for any vector $\boldsymbol{b}$ with $\boldsymbol{b} \leqslant \boldsymbol{a}/2$, one has
$$\boldsymbol{x}^{\boldsymbol{a}} = \left(\boldsymbol{x}^{\boldsymbol{b}} \cdot \boldsymbol{x}^{\boldsymbol{a}/2-\boldsymbol{b}}\right)^2 \geqslant \left(\min\{\boldsymbol{x}^{\boldsymbol{b}}, \boldsymbol{x}^{\boldsymbol{a}/2-\boldsymbol{b}}\}\right)^4.$$
We therefore deduce that either $\boldsymbol{b} \in S(\boldsymbol{a};\boldsymbol{x})$, or $\boldsymbol{a}/2 - \boldsymbol{b} \in S(\boldsymbol{a};\boldsymbol{x})$. For each vector $\boldsymbol{b}$ with $\boldsymbol{b} \leqslant \boldsymbol{a}/2$, define the function $\phi(\boldsymbol{b})$ by taking $\phi(\boldsymbol{b}) = \boldsymbol{b}$ or $\phi(\boldsymbol{b}) = \boldsymbol{a}/2 - \boldsymbol{b}$, in such a manner that $\phi(\boldsymbol{b}) \in S(\boldsymbol{a};\boldsymbol{x})$. We note that whenever $\phi(\boldsymbol{b}) = \phi(\boldsymbol{b}')$, then necessarily $\boldsymbol{b} = \boldsymbol{b}'$ or $\boldsymbol{b} = \boldsymbol{a}/2 - \boldsymbol{b}'$. In particular, as we consider the vectors $\phi(\boldsymbol{b})$ as $\boldsymbol{b}$ varies over all vectors with $\boldsymbol{b} \leqslant \boldsymbol{a}/2$, one finds that each value $\phi(\boldsymbol{b})$ appears at most twice. We therefore deduce that
$$G(\boldsymbol{a};\boldsymbol{x}) \geqslant \frac{1}{2} \sum_{\boldsymbol{b} \leqslant \boldsymbol{a}/2} \widetilde{g}(\phi(\boldsymbol{b}))$$
$$\geqslant \frac{1}{2} \sum_{\boldsymbol{b} \leqslant \boldsymbol{a}/2} \min\{\widetilde{g}(\boldsymbol{b}), \widetilde{g}(\boldsymbol{a}/2 - \boldsymbol{b})\} = f(\boldsymbol{a}),$$
whence $C(\boldsymbol{a};\boldsymbol{x}) \leqslant F(\boldsymbol{a})$ for every vector $\boldsymbol{x} \in \mathbb{N}^k$. The conclusion of the lemma follows at once. □

The next lemma permits substantial simplifications to be made in this phase of our argument.

LEMMA 9.5. — *Whenever $\boldsymbol{a}, \boldsymbol{a}' \in V_2'$, one has*
$$F(\boldsymbol{a}, \boldsymbol{a}') \leqslant \frac{1}{2} F(\boldsymbol{a}) F(\boldsymbol{a}').$$

Proof. — If either $\boldsymbol{a}$ or $\boldsymbol{a}'$ is empty, then the conclusion of the lemma is immediate from our convention that $F(\mathfrak{z}) = 2$. Suppose then that neither $\boldsymbol{a}$ nor $\boldsymbol{a}'$ is empty. Given vectors $\boldsymbol{b}$ and $\boldsymbol{b}'$, it is evident that $(\boldsymbol{b}, \boldsymbol{b}') \leqslant (\boldsymbol{a}, \boldsymbol{a}')/2$ if and only if $\boldsymbol{b} \leqslant \boldsymbol{a}/2$ and $\boldsymbol{b}' \leqslant \boldsymbol{a}'/2$. We therefore obtain
$$f(\boldsymbol{a}, \boldsymbol{a}') = \frac{1}{2} \sum_{\boldsymbol{b} \leqslant \boldsymbol{a}/2} \sum_{\boldsymbol{b}' \leqslant \boldsymbol{a}'/2} \min\{\widetilde{g}(\boldsymbol{b})\widetilde{g}(\boldsymbol{b}'), \widetilde{g}(\boldsymbol{a}/2 - \boldsymbol{b})\widetilde{g}(\boldsymbol{a}'/2 - \boldsymbol{b}')\}$$
$$\geqslant \frac{1}{2} \sum_{\boldsymbol{b} \leqslant \boldsymbol{a}/2} \sum_{\boldsymbol{b}' \leqslant \boldsymbol{a}'/2} \min\{\widetilde{g}(\boldsymbol{b}), \widetilde{g}(\boldsymbol{a}/2 - \boldsymbol{b})\} \min\{\widetilde{g}(\boldsymbol{b}'), \widetilde{g}(\boldsymbol{a}'/2 - \boldsymbol{b}')\}$$
$$= 2 f(\boldsymbol{a}) f(\boldsymbol{a}'),$$
and the desired conclusion follows from (9.7) and the definition of the function F. □

LEMMA 9.6. — *Suppose that $\boldsymbol{a} \in V_2'$, and denote by s the number of components of $\boldsymbol{a}$ that are equal to 2. The following upper bounds hold.*

(i) When $s = 0$, one has $F(\boldsymbol{a}) \leqslant 2$.
(ii) When s is even, one has $F(\boldsymbol{a}) \leqslant 9/4$.
(iii) In any case, one has $F(\boldsymbol{a}) \leqslant 3$.

Proof. — By the definitions of the functions f and F, one finds that
$$f(4) = 5/2, \quad f(6) = 4, \quad f(10) = 7,$$
and hence
$$F(4) = 2, \quad F(6) = 7/4 < 2, \quad F(10) = 11/7 < 2. \tag{9.9}$$
When $s = 0$, therefore, repeated application of Lemma 9.5 in combination with the bounds (9.9) demonstrates that $F(\boldsymbol{a}) \leqslant 2$, the latter conclusion being trivial when $\boldsymbol{a} = \mathfrak{z}$.

Suppose next that $\boldsymbol{a} \in V_2'$ and that s is positive. By repeated application of Lemma 9.5 in concert with (9.9), we may delete any components of $\boldsymbol{a}$ that are not equal to 2 without increasing the value of $F(\boldsymbol{a})$. We may consequently suppose without loss of generality that $\boldsymbol{a} = \boldsymbol{2}_s$, where we write $\boldsymbol{2}_s$ for the s-dimensional vector $(2, 2, \ldots, 2)$. But $f(\boldsymbol{2}_s)$ may be computed explicitly via an elementary combinatorial argument. Indeed, it is an easy exercise to confirm that when $s = 2m - 1$ with $m \in \mathbb{N}$, one has
$$f(\boldsymbol{2}_s) = \sum_{j=0}^{m-1} \binom{2m-1}{j} 3^j,$$
and that when $s = 2m$ with $m \in \mathbb{N}$, then
$$f(\boldsymbol{2}_s) = \frac{1}{2}\binom{2m}{m} 3^m + \sum_{j=0}^{m-1} \binom{2m}{j} 3^j.$$
In order to simplify our discussion of the numbers $F(\boldsymbol{2}_s)$, we note next that since $F(\boldsymbol{2}_s) = 3^s/f(\boldsymbol{2}_s)$, we have for each natural number m the identities
$$F(\boldsymbol{2}_{2m+1})^{-1} - F(\boldsymbol{2}_{2m-1})^{-1} = 3^{-2m-1} + \sum_{j=0}^{m-1} \left(\binom{2m+1}{j+1} - 3\binom{2m-1}{j}\right) 3^{j-2m},$$
and
$$F(\boldsymbol{2}_{2m+2})^{-1} - F(\boldsymbol{2}_{2m})^{-1} = 3^{-2m-2} + \sum_{j=0}^{m-1} \left(\binom{2m+2}{j+1} - 3\binom{2m}{j}\right) 3^{j-2m-1}$$
$$+ \frac{1}{2}\left(\binom{2m+2}{m+1} - 3\binom{2m}{m}\right) 3^{-m-1}.$$
But whenever $0 \leqslant j \leqslant s/2$, one has
$$\binom{s+2}{j+1}\binom{s}{j}^{-1} = \frac{4(s+2)(s+1)}{(s+2)^2 - (2j-s)^2} \geqslant 3,$$
and thus we recognise that both of the sequences $\{F(\boldsymbol{2}_{2m-1})\}$ and $\{F(\boldsymbol{2}_{2m})\}$ are monotone decreasing with m. In particular, one has $F(\boldsymbol{2}_s) \leqslant F(\boldsymbol{2}_1) = 3$ when s is

odd, and $F(\mathbf{2}_s) \leqslant F(\mathbf{2}_2) = 9/4$ when s is even. This establishes parts (ii) and (iii) of the lemma, and thereby completes its proof. $\square$

The second phase of our argument is now complete. Lemma 9.6 shows that $F(\boldsymbol{a}) \leqslant 3$ for every $\boldsymbol{a} \in V_2'$, whence it follows from Lemma 9.4 that $C(\boldsymbol{a}) \leqslant 3$ for every $\boldsymbol{a} \in V_2$. Since $C(1,1,1) = 8$, it follows that the elements of V_2 play no role in determining the supremum appearing in (9.5). It remains only to investigate the vectors $\boldsymbol{a} \in V_1$ having one or more components equal to 1. This phase of our argument is the most difficult yet, and requires a third triumvirate of lemmata. We begin with some additional notation of somewhat peculiar flavour. When $\boldsymbol{a}$ is a k-dimensional vector in V, we refer to a set $A = \{\boldsymbol{a}_1, \boldsymbol{a}_2, \boldsymbol{a}_3, \boldsymbol{a}_4\}$ of four vectors in $(\mathbb{N} \cup \{0\})^k$ as a *decomposition* of $\boldsymbol{a}$, when the four vectors in A are pairwise distinct and

$$\boldsymbol{a}_1 + \boldsymbol{a}_2 + \boldsymbol{a}_3 + \boldsymbol{a}_4 = \boldsymbol{a}. \tag{9.10}$$

When $\widetilde{g}(\boldsymbol{a}_j) \geqslant l$ for $1 \leqslant j \leqslant 4$, we describe such a decomposition A as an *l-decomposition* of $\boldsymbol{a}$. Finally, we describe a set $\mathcal{A}$ as a (ν, μ)-*set of l-decompositions* of $\boldsymbol{a}$, when $\mathcal{A}$ consists of ν l-decompositions of $\boldsymbol{a}$ and for every $\boldsymbol{b} \in (\mathbb{N} \cup \{0\})^k$, one has

$$\operatorname{card}\{A \in \mathcal{A} : \boldsymbol{b} \in A\} \leqslant \mu. \tag{9.11}$$

LEMMA 9.7. — *Suppose that $\boldsymbol{a} \in V$, and that there exists a (ν, μ)-set of l-decompositions of $\boldsymbol{a}$. Then whenever $\boldsymbol{a}' \in V_2'$, one has*

$$C(\boldsymbol{a}, \boldsymbol{a}') \leqslant \frac{\widetilde{\tau}(\boldsymbol{a})}{\lceil 2\nu/\mu \rceil l} F(\boldsymbol{a}').$$

Proof. — Suppose that $\boldsymbol{a} \in V$ is a k-dimensional vector, and let $\mathcal{A}$ be a (ν, μ)-set of l-decompositions of $\boldsymbol{a}$. When $A = \{\boldsymbol{a}_1, \ldots, \boldsymbol{a}_4\} \in \mathcal{A}$ and $\boldsymbol{x} \in \mathbb{N}^k$, it follows from the definition (9.10) of a decomposition that

$$\boldsymbol{x}^{\boldsymbol{a}_1} \cdot \boldsymbol{x}^{\boldsymbol{a}_2} \cdot \boldsymbol{x}^{\boldsymbol{a}_3} \cdot \boldsymbol{x}^{\boldsymbol{a}_4} = \boldsymbol{x}^{\boldsymbol{a}},$$

whence $\boldsymbol{x}^{4\boldsymbol{a}_j} \leqslant \boldsymbol{x}^{\boldsymbol{a}}$ for some $\boldsymbol{a}_j \in A$. We therefore deduce that $\boldsymbol{a}_j \in S(\boldsymbol{a};\boldsymbol{x})$ for some $\boldsymbol{a}_j \in A$. We note here that since $A \in \mathcal{A}$ is an l-decomposition, then $\widetilde{g}(\boldsymbol{a}_j) \geqslant l$. Since $\mathcal{A}$ is a (ν, μ)-set of l-decompositions, we obtain ν vectors $\boldsymbol{a}_j \in S(\boldsymbol{a};\boldsymbol{x})$ on considering each of the ν l-decompositions A of $\boldsymbol{a}$. In view of (9.11), it follows from the pigeon-hole principle that at least $\lceil \nu/\mu \rceil$ of the latter vectors are pairwise distinct. Henceforth, we denote this set of $\lceil \nu/\mu \rceil$ distinct vectors in $S(\boldsymbol{a};\boldsymbol{x})$ by $\langle \mathcal{A}; \boldsymbol{x} \rangle$.

Equipped with the preliminary observations of the previous paragraph, the proof of the lemma is rapidly completed in the case that $\boldsymbol{a}' = \boldsymbol{\mathfrak{z}}$. On considering the contribution to $G(\boldsymbol{a};\boldsymbol{x})$ arising from the vectors in $\langle \mathcal{A}; \boldsymbol{x} \rangle$, we find that

$$G(\boldsymbol{a};\boldsymbol{x}) \geqslant \sum_{\boldsymbol{c} \in \langle \mathcal{A};\boldsymbol{x}\rangle} \widetilde{g}(\boldsymbol{c}) \geqslant \lceil \nu/\mu \rceil l.$$

But for any real number z, one plainly has

$$2\lceil z \rceil \geqslant \lceil 2z \rceil. \tag{9.12}$$

Thus we conclude that for any $\boldsymbol{x} \in \mathbb{N}^k$, one has
$$C(\boldsymbol{a}; \boldsymbol{x}) \leqslant \frac{2\widetilde{\tau}(\boldsymbol{a})}{\lceil 2\nu/\mu \rceil l} = \frac{\widetilde{\tau}(\boldsymbol{a})}{\lceil 2\nu/\mu \rceil l} F(\mathfrak{z}).$$
The conclusion of the lemma is therefore immediate when $\boldsymbol{a}' = \mathfrak{z}$.

Suppose next that $\boldsymbol{a}'$ is a k'-dimensional vector in V_2. Let $A = \{\boldsymbol{a}_1, \ldots, \boldsymbol{a}_4\}$ be an l-decomposition of $\boldsymbol{a}$ belonging to $\mathcal{A}$, and let $\boldsymbol{b}$ be a vector with $\boldsymbol{b} \leqslant \boldsymbol{a}'/2$. We observe that for any permutation σ of the set $\{1, 2, 3, 4\}$, the set of vectors
$$\{(\boldsymbol{a}_{\sigma(1)}, \boldsymbol{b}), (\boldsymbol{a}_{\sigma(2)}, \boldsymbol{b}), (\boldsymbol{a}_{\sigma(3)}, \boldsymbol{a}'/2 - \boldsymbol{b}), (\boldsymbol{a}_{\sigma(4)}, \boldsymbol{a}'/2 - \boldsymbol{b})\}$$
forms an $L(\boldsymbol{b})$-decomposition of $(\boldsymbol{a}, \boldsymbol{a}')$, where
$$L(\boldsymbol{b}) = l \cdot \min\{\widetilde{g}(\boldsymbol{b}), \widetilde{g}(\boldsymbol{a}'/2 - \boldsymbol{b})\}. \tag{9.13}$$
We denote by $A(\boldsymbol{b})$ the set of such $L(\boldsymbol{b})$-decompositions of $(\boldsymbol{a}, \boldsymbol{a}')$ as σ runs over the permutations of $\{1, 2, 3, 4\}$. Our argument divides naturally according to whether or not $\boldsymbol{b} = \boldsymbol{a}'/2 - \boldsymbol{b}$, and we simplify our account by defining parameters $\widetilde{\nu} = \widetilde{\nu}(\boldsymbol{b})$ and $\widetilde{\mu} = \widetilde{\mu}(\boldsymbol{b})$ by
$$(\widetilde{\nu}, \widetilde{\mu}) = \begin{cases} (6, 3), & \text{when } 4\boldsymbol{b} \neq \boldsymbol{a}', \\ (1, 1), & \text{when } 4\boldsymbol{b} = \boldsymbol{a}'. \end{cases}$$
It then follows that $A(\boldsymbol{b})$ is a $(\widetilde{\nu}, \widetilde{\mu})$-set of $L(\boldsymbol{b})$-decompositions of $(\boldsymbol{a}, \boldsymbol{a}')$. Moreover, if we denote by $\mathcal{A}(\boldsymbol{b})$ the union of the sets $A(\boldsymbol{b})$ for $A \in \mathcal{A}$, we see that $\mathcal{A}(\boldsymbol{b})$ is a $(\nu\widetilde{\nu}, \mu\widetilde{\mu})$-set of $L(\boldsymbol{b})$-decompositions of $(\boldsymbol{a}, \boldsymbol{a}')$.

We now imitate the argument applied earlier in the case where $\boldsymbol{a}' = \mathfrak{z}$, constructing for each vector $\boldsymbol{b}$ with $\boldsymbol{b} \leqslant \boldsymbol{a}'/2$, and each $\boldsymbol{x} \in \mathbb{N}^{k+k'}$, a set $\langle \mathcal{A}(\boldsymbol{b}), \boldsymbol{x} \rangle$. As in our previous discussion, this set $\langle \mathcal{A}(\boldsymbol{b}), \boldsymbol{x} \rangle$ consists of $\lceil \widetilde{\nu}\nu/(\widetilde{\mu}\mu) \rceil$ distinct vectors in $S((\boldsymbol{a}, \boldsymbol{a}'); \boldsymbol{x})$, each of which is picked up from an $L(\boldsymbol{b})$-decomposition in $\mathcal{A}(\boldsymbol{b})$. We therefore obtain the lower bound
$$\sum_{\boldsymbol{c} \in \langle \mathcal{A}(\boldsymbol{b}), \boldsymbol{x} \rangle} \widetilde{g}(\boldsymbol{c}) \geqslant \lceil \widetilde{\nu}\nu/(\widetilde{\mu}\mu) \rceil L(\boldsymbol{b}). \tag{9.14}$$
In order to assess the total contribution to $G((\boldsymbol{a}, \boldsymbol{a}'); \boldsymbol{x})$ arising from vectors of this type, we put
$$I = \bigcup_{\boldsymbol{b} \leqslant \boldsymbol{a}'/2} \langle \mathcal{A}(\boldsymbol{b}), \boldsymbol{x} \rangle,$$
and note that $I \subseteq S((\boldsymbol{a}, \boldsymbol{a}'); \boldsymbol{x})$. Then on observing that $\langle \mathcal{A}(\boldsymbol{b}), \boldsymbol{x} \rangle$ and $\langle \mathcal{A}(\boldsymbol{b}'), \boldsymbol{x} \rangle$ are disjoint unless $\boldsymbol{b}' = \boldsymbol{b}$ or $\boldsymbol{b}' = \boldsymbol{a}'/2 - \boldsymbol{b}$, we find that
$$G((\boldsymbol{a}, \boldsymbol{a}'); \boldsymbol{x}) \geqslant \sum_{\boldsymbol{c} \in I} \widetilde{g}(\boldsymbol{c}) \geqslant \sum_{\boldsymbol{b} \leqslant \boldsymbol{a}'/2} \frac{1}{\delta(\boldsymbol{b})} \sum_{\boldsymbol{c} \in \langle \mathcal{A}(\boldsymbol{b}), \boldsymbol{x} \rangle} \widetilde{g}(\boldsymbol{c}), \tag{9.15}$$
where
$$\delta(\boldsymbol{b}) = \begin{cases} 2, & \text{when } 4\boldsymbol{b} \neq \boldsymbol{a}', \\ 1, & \text{when } 4\boldsymbol{b} = \boldsymbol{a}'. \end{cases} \tag{9.16}$$

The latter relation ensures that $\delta(\boldsymbol{b}) = \widetilde{\nu}/\widetilde{\mu}$ for every vector $\boldsymbol{b}$, so that on collecting together (9.13)-(9.15), and making use again of (9.12), we deduce that

$$G((\boldsymbol{a},\boldsymbol{a}');\boldsymbol{x}) \geqslant \frac{1}{2} \sum_{\boldsymbol{b} \leqslant \boldsymbol{a}'/2} \lceil 2\nu/\mu \rceil l \min\{\widetilde{g}(\boldsymbol{b}), \widetilde{g}(\boldsymbol{a}'/2 - \boldsymbol{b})\}$$
$$= l \lceil 2\nu/\mu \rceil f(\boldsymbol{a}').$$

On recalling (9.7), we therefore conclude that for every $\boldsymbol{x} \in \mathbb{N}^{k+k'}$, one has

$$C((\boldsymbol{a},\boldsymbol{a}');\boldsymbol{x}) \leqslant \frac{\widetilde{\tau}(\boldsymbol{a},\boldsymbol{a}')}{l \lceil 2\nu/\mu \rceil f(\boldsymbol{a}')} = \frac{\widetilde{\tau}(\boldsymbol{a})}{l \lceil 2\nu/\mu \rceil} F(\boldsymbol{a}').$$

The conclusion of the lemma now follows immediately whenever $\boldsymbol{a}' \in V_2$. $\square$

We also require a variant of the previous lemma that can, however, be established with a similar argument.

LEMMA 9.8. — *Suppose that $\boldsymbol{a} \in V$, and that for each j with $1 \leqslant j \leqslant J$, there exists an l_j-decomposition of $\boldsymbol{a}$, say A_j. Suppose also that the A_j, for $1 \leqslant j \leqslant J$, are pairwise disjoint. Then whenever $\boldsymbol{a}' \in V_2'$, one has*

$$C(\boldsymbol{a},\boldsymbol{a}') \leqslant \frac{1}{2}\widetilde{\tau}(\boldsymbol{a})\Big(\sum_{j=1}^{J} l_j\Big)^{-1} F(\boldsymbol{a}').$$

Proof. — Suppose that $\boldsymbol{a} \in V$ is a k-dimensional vector. As in the opening discussion of the proof of the previous lemma, for each $\boldsymbol{x} \in \mathbb{N}^k$, and for each index j with $1 \leqslant j \leqslant J$, one may choose a vector $\boldsymbol{c}_j$ from A_j with $\boldsymbol{c}_j \in S(\boldsymbol{a};\boldsymbol{x})$. Since the A_j are pairwise disjoint, the vectors $\boldsymbol{c}_j$ are automatically pairwise distinct for $1 \leqslant j \leqslant J$. Also, since each A_j is an l_j-decomposition, one has $\widetilde{g}(\boldsymbol{c}_j) \geqslant l_j$ ($1 \leqslant j \leqslant J$). We therefore deduce that for each $\boldsymbol{x} \in \mathbb{N}^k$,

$$G(\boldsymbol{a};\boldsymbol{x}) \geqslant \sum_{j=1}^{J} \widetilde{g}(\boldsymbol{c}_j) \geqslant \sum_{j=1}^{J} l_j,$$

and the conclusion of the lemma swiftly follows when $\boldsymbol{a}' = \mathfrak{z}$, on recalling that $F(\mathfrak{z}) = 2$.

Suppose next that $\boldsymbol{a}'$ is a k'-dimensional vector in V_2. Adopting the notation of the proof of Lemma 9.7, and imitating the argument leading to the bound (9.14), for every vector $\boldsymbol{b}$ with $\boldsymbol{b} \leqslant \boldsymbol{a}'/2$, and for each A_j with $1 \leqslant j \leqslant J$, we construct a set $A_j(\boldsymbol{b})$. In the present situation, the latter object is a $(\widetilde{\nu}, \widetilde{\mu})$-set of $L_j(\boldsymbol{b})$-decompositions of $(\boldsymbol{a},\boldsymbol{a}')$, where

$$L_j(\boldsymbol{b}) = l_j \cdot \min\{\widetilde{g}(\boldsymbol{b}), \widetilde{g}(\boldsymbol{a}'/2 - \boldsymbol{b})\}.$$

Given a vector $\boldsymbol{x} \in \mathbb{N}^{k+k'}$, we next construct a set $\langle A_j(\boldsymbol{b});\boldsymbol{x}\rangle$, every member of which is chosen from an $L_j(\boldsymbol{b})$-decomposition in $A_j(\boldsymbol{b})$, and belongs to $S((\boldsymbol{a},\boldsymbol{a}');\boldsymbol{x})$. As before, the cardinality of $\langle \mathcal{A}_j(\boldsymbol{b}); \boldsymbol{x}\rangle$ is at least $\widetilde{\nu}/\widetilde{\mu} = \delta(\boldsymbol{b})$. Note also that $\langle \mathcal{A}_j(\boldsymbol{b}); \boldsymbol{x}\rangle$ and $\langle \mathcal{A}_j(\boldsymbol{b}'); \boldsymbol{x}\rangle$ are disjoint unless $\boldsymbol{b}' = \boldsymbol{b}$ or $\boldsymbol{b}' = \boldsymbol{a}'/2 - \boldsymbol{b}$. On writing

$$I_j = \bigcup_{\boldsymbol{b} \leqslant \boldsymbol{a}'/2} \langle \mathcal{A}_j(\boldsymbol{b}); \boldsymbol{x}\rangle,$$

we therefore deduce that

$$\sum_{\boldsymbol{c}\in I_j} \widetilde{g}(\boldsymbol{c}) \geqslant \sum_{\boldsymbol{b}\leqslant \boldsymbol{a}'/2} \frac{1}{\delta(\boldsymbol{b})} \sum_{\boldsymbol{c}\in \langle \mathcal{A}_j(\boldsymbol{b}); \boldsymbol{x}\rangle} \widetilde{g}(\boldsymbol{c}) \geqslant \sum_{\boldsymbol{b}\leqslant \boldsymbol{a}'/2} \frac{1}{\delta(\boldsymbol{b})} \left\lceil \frac{\widetilde{\nu}}{\widetilde{\mu}}\right\rceil L_j(\boldsymbol{b})$$

$$= \sum_{\boldsymbol{b}\leqslant \boldsymbol{a}'/2} l_j \min\{\widetilde{g}(\boldsymbol{b}), \widetilde{g}(\boldsymbol{a}'/2 - \boldsymbol{b})\} = 2l_j f(\boldsymbol{a}').$$

But $I_j \subseteq S((\boldsymbol{a},\boldsymbol{a}'); \boldsymbol{x})$ for $1 \leqslant j \leqslant J$, and moreover I_j and $I_{j'}$ are disjoint whenever $j \neq j'$. Thus we arrive at the lower bound

$$G((\boldsymbol{a},\boldsymbol{a}'); \boldsymbol{x}) \geqslant \sum_{j=1}^J \sum_{\boldsymbol{c}\in I_j} \widetilde{g}(\boldsymbol{c}) \geqslant 2f(\boldsymbol{a}') \sum_{j=1}^J l_j,$$

whence for every $\boldsymbol{x} \in \mathbb{N}^{k+k'}$, we derive the upper bound

$$C((\boldsymbol{a},\boldsymbol{a}'); \boldsymbol{x}) \leqslant \frac{\widetilde{\tau}(\boldsymbol{a},\boldsymbol{a}')}{2f(\boldsymbol{a}')\sum_{j=1}^J l_j} = \frac{1}{2}\widetilde{\tau}(\boldsymbol{a})\Big(\sum_{j=1}^J l_j\Big)^{-1} F(\boldsymbol{a}').$$

The desired conclusion now follows immediately for each $\boldsymbol{a}' \in V_2$. □

Our careful preparations now complete, we at last launch our assault on the evaluation of $C(\boldsymbol{a})$. We must nonetheless exhibit some fortitude if we are to successfully storm the citadel.

LEMMA 9.9. — *Whenever $\boldsymbol{a} \in V_1$, one has $C(\boldsymbol{a}) \leqslant 8$.*

Proof. — Our argument splits into cases according to the number of components of the vector in question that are equal to 1 or 2. Given a vector $\boldsymbol{b}$ (possibly empty), we denote by $t(\boldsymbol{b})$ the number of components of $\boldsymbol{b}$ equal to 1, and by $s(\boldsymbol{b})$ the number of components equal to 2. Consider a fixed vector $\boldsymbol{a} \in V_1$, and recall throughout that we may permute the components of $\boldsymbol{a}$ with impunity, whenever we are so-inclined.

(i) Suppose that $t(\boldsymbol{a}) = 0$. In this situation one has $\boldsymbol{a} \in V_2$, and as we have already discussed, the conclusion of the lemma is then immediate from Lemmata 9.4 and 9.6.

(ii) Suppose that $t(\boldsymbol{a}) = 1$. When $\boldsymbol{a} = (1)$, the trivial bound (9.4) yields $C(1) \leqslant 2$, which suffices for our purpose. When $\boldsymbol{a} = (1, \boldsymbol{a}')$ with $\boldsymbol{a}' \in V_2$, on the other hand, one may apply Lemmata 9.1, 9.4 and 9.6 (iii) in concert with our previous estimate to deduce that

$$C(\boldsymbol{a}) \leqslant C(1)C(\boldsymbol{a}') \leqslant 2 \times 3 = 6.$$

(iii) Suppose that $t(\boldsymbol{a}) = 2$. We now subdivide our argument according to the value of $s(\boldsymbol{a})$.

When $s(\boldsymbol{a}) = 0$, we may suppose that $\boldsymbol{a}$ takes the shape $\boldsymbol{a} = (1, 1, \boldsymbol{a}')$ with $\boldsymbol{a}' \in V_2'$ and $s(\boldsymbol{a}') = 0$. The trivial bound (9.4) now yields $C(1,1) \leqslant 4$. Meanwhile, when $\boldsymbol{a}' \in V_2$ satisfies $s(\boldsymbol{a}') = 0$, one finds that Lemmata 9.1, 9.4 and 9.6(i) lead from the last bound to the estimate

$$C(1, 1, \boldsymbol{a}') \leqslant C(1,1)C(\boldsymbol{a}') \leqslant 4 \times 2 = 8.$$

Suppose next that $s(\boldsymbol{a}) = 1$. When $\boldsymbol{a} = (1, 1, 2)$, we observe that for each $\boldsymbol{x} \in \mathbb{N}^3$, one has

$$\boldsymbol{x}^{\boldsymbol{a}} = \boldsymbol{x}^{(1,0,0)} \boldsymbol{x}^{(0,1,0)} \boldsymbol{x}^{(0,0,1)} \boldsymbol{x}^{(0,0,1)}.$$

It follows that the set $S(\boldsymbol{a}; \boldsymbol{x})$ contains $(0,0,0)$ together with one at least of $(1,0,0)$, $(0,1,0)$ and $(0,0,1)$. We therefore obtain $G(\boldsymbol{a}; \boldsymbol{x}) \geqslant 1 + 3 = 4$, whence for each $\boldsymbol{x} \in \mathbb{N}^3$ we have $C(\boldsymbol{a}; \boldsymbol{x}) \leqslant 12/4 = 3$. The bound $C(1,1,2) \leqslant 3$ follows immediately. When $\boldsymbol{a} = (1, 1, 2, \boldsymbol{a}')$ with $\boldsymbol{a}' \in V_2$, meanwhile, our hypothesis that $s(\boldsymbol{a}) = 1$ implies that $s(\boldsymbol{a}') = 0$. In this case we may combine the conclusion just obtained with Lemmata 9.1, 9.4 and 9.6 (i) to deduce that

$$C(\boldsymbol{a}) \leqslant C(1,1,2)C(\boldsymbol{a}') \leqslant 3 \times 2 = 6.$$

Finally, suppose that $s(\boldsymbol{a}) \geqslant 2$. In this case we may suppose that $\boldsymbol{a} = (1, 1, 2, 2, \boldsymbol{a}')$ with $\boldsymbol{a}' \in V_2'$. Putting

$$A_1 = \{(1,0,0,0), (0,1,0,0), (0,0,2,0), (0,0,0,2)\},$$
$$A_2 = \{(1,0,1,0), (0,1,0,1), (0,0,1,0), (0,0,0,1)\},$$
$$A_3 = \{(1,0,0,1), (0,1,1,0), (0,0,1,1), (0,0,0,0)\},$$

one may verify that A_1 and A_2 are 3-decompositions of $(1,1,2,2)$, whilst A_3 is a 1-decomposition of $(1,1,2,2)$. Noting also that A_1, A_2 and A_3 are pairwise disjoint, we conclude from Lemmata 9.6 (iii) and 9.8 that

$$C(\boldsymbol{a}) = C(1, 1, 2, 2, \boldsymbol{a}') \leqslant \frac{\widetilde{\tau}(1,1,2,2)}{3+3+1} \cdot \frac{F(\boldsymbol{a}')}{2} \leqslant \frac{36}{7} \cdot \frac{3}{2} = \frac{54}{7}.$$

Collecting together the above estimates, we find that $C(\boldsymbol{a}) \leqslant 8$ whenever $t(\boldsymbol{a}) = 2$.

(iv) Suppose that $t(\boldsymbol{a}) = 3$. Again we subdivide our argument according to the value of $s(\boldsymbol{a})$.

When $s(\boldsymbol{a}) = 0$, we may suppose that $\boldsymbol{a} = (1, 1, 1, \boldsymbol{a}')$ with $\boldsymbol{a}' \in V_2'$ and $s(\boldsymbol{a}') = 0$. Since the set

$$\{(0,0,0), (1,0,0), (0,1,0), (0,0,1)\}$$

provides a 1-decomposition of $(1,1,1)$, Lemmata 9.6 (i) and 9.8 yield

$$C(\boldsymbol{a}) = C(1, 1, 1, \boldsymbol{a}') \leqslant \frac{\widetilde{\tau}(1,1,1)}{1} \cdot \frac{F(\boldsymbol{a}')}{2} \leqslant 8.$$

Suppose next that $s(\boldsymbol{a})$ is odd. In this case we may suppose that $\boldsymbol{a} = (1,1,1,2,\boldsymbol{a}')$ with $\boldsymbol{a}' \in V_2'$ and $s(\boldsymbol{a}')$ even. On observing that the two disjoint sets

$$\{(1,0,0,0), (0,1,0,0), (0,0,1,0), (0,0,0,2)\},$$
$$\{(1,1,0,0), (0,0,1,1), (0,0,0,1), (0,0,0,0)\}$$

are, respectively, 3- and 1-decompositions of $(1,1,1,2)$, we deduce from Lemmata 9.6 (ii) and 9.8 that

$$C(\boldsymbol{a}) \leqslant \frac{\widetilde{\tau}(1,1,1,2)}{3+1} \cdot \frac{F(\boldsymbol{a}')}{2} \leqslant \frac{24}{4} \cdot \frac{9/4}{2} = \frac{27}{4}.$$

Finally, suppose that $s(\boldsymbol{a})$ is a positive even number. We may now suppose that $\boldsymbol{a} = (1,1,1,2,2,\boldsymbol{a}')$ with $\boldsymbol{a}' \in V_2'$ and $s(\boldsymbol{a}')$ even. In this situation, we note that the four disjoint sets

$$\{(1,0,0,0,0), (0,1,0,0,0), (0,0,1,1,1), (0,0,0,1,1)\},$$
$$\{(1,0,0,1,0), (0,1,0,1,0), (0,0,1,0,1), (0,0,0,0,1)\},$$
$$\{(1,0,0,0,1), (0,1,0,0,1), (0,0,1,1,0), (0,0,0,1,0)\},$$
$$\{(1,1,0,0,0), (0,0,1,0,0), (0,0,0,2,0), (0,0,0,0,2)\}$$

are 3-decompositions of $(1,1,1,2,2)$. Thus we find that Lemmata 9.6 (ii) and 9.8 in this case yield the estimate

$$C(\boldsymbol{a}) \leqslant \frac{\widetilde{\tau}(1,1,1,2,2)}{3+3+3+3} \cdot \frac{F(\boldsymbol{a}')}{2} \leqslant \frac{72}{12} \cdot \frac{9/4}{2} = \frac{27}{4}.$$

Collecting together the above estimates, we find that $C(\boldsymbol{a}) \leqslant 8$ whenever $t(\boldsymbol{a}) = 3$.

(v) Suppose that $t(\boldsymbol{a}) = 4m$ with $m \geqslant 1$. In order to simplify our argument, we now introduce the notation of writing $\mathbf{1}_t$ for the t-dimensional vector $(1,1,\ldots,1)$. Thus, in the situation presently under consideration, we may write $\boldsymbol{a} = (\mathbf{1}_{4m}, \boldsymbol{a}')$ with $\boldsymbol{a}' \in V_2'$.

We consider the set $\mathcal{A}_{4m}$ of all the 3^m-decompositions of $\mathbf{1}_{4m}$. Notice that whenever $A = \{\boldsymbol{a}_1, \ldots, \boldsymbol{a}_4\}$ is a 3^m-decomposition of $\mathbf{1}_{4m}$, then $\boldsymbol{a}_j \in \{0,1\}^{4m}$ and $t(\boldsymbol{a}_j) \geqslant m$ for $1 \leqslant j \leqslant 4$, and furthermore $\boldsymbol{a}_1 + \cdots + \boldsymbol{a}_4 = \mathbf{1}_{4m}$. Plainly, therefore, one must have $t(\boldsymbol{a}_j) = m$ for $1 \leqslant j \leqslant 4$, and an elementary combinatorial argument leads to the conclusion that $\mathcal{A}_{4m}$ is a (ν_{4m}, μ_{4m})-set of 3^m-decompositions of $\mathbf{1}_{4m}$, where

$$\nu_{4m} = \frac{(4m)!}{4!(m!)^4} \quad \text{and} \quad \mu_{4m} = \frac{(3m)!}{3!(m!)^3}.$$

Consequently, Lemmata 9.6 (iii) and 9.7 yield the bound

$$C(\boldsymbol{a}) = C(\mathbf{1}_{4m}, \boldsymbol{a}') \leqslant \frac{\widetilde{\tau}(\mathbf{1}_{4m})}{2\nu_{4m}/\mu_{4m}} \cdot \frac{F(\boldsymbol{a}')}{3^m} \leqslant 2^{4m+1} 3^{1-m} \binom{4m}{m}^{-1}.$$

But when $m \geqslant 1$, one has $\binom{4m+4}{m+1} = 4\omega(m)\binom{4m}{m}$, where

$$\omega(m) = \frac{(4m+1)(4m+2)(4m+3)}{(3m+1)(3m+2)(3m+3)}.$$

MÉMOIRES DE LA SMF 100

CHAPTER 9. AN AUXILIARY BOUND FOR THE DIVISOR FUNCTION

On noting that $\omega(1) = 7/4$ and $\omega(m+1) \geqslant \omega(m)$ for $m \geqslant 1$, we deduce that for $m \geqslant 1$, one has
$$\binom{4m}{m} \geqslant 4 \cdot 7^{m-1}, \tag{9.17}$$
whence
$$C(\boldsymbol{a}) \leqslant 2^{4m+1} 3^{1-m} (4 \cdot 7^{m-1})^{-1} = 8 \cdot \left(\frac{16}{21}\right)^{m-1} \leqslant 8.$$

(vi) Suppose that $t(\boldsymbol{a}) = 4m+1$ with $m \geqslant 1$. We may now write $\boldsymbol{a} = (\mathbf{1}_{4m+1}, \boldsymbol{a}')$ with $\boldsymbol{a}' \in V_2'$. Following the argument of the previous case, we consider the set $\mathcal{A}_{4m+1}$ of all the 3^m-decompositions of $\mathbf{1}_{4m+1}$. When $A \in \mathcal{A}_{4m+1}$, we may now suppose that
$$A = \{\boldsymbol{a}_1, \ldots, \boldsymbol{a}_4\}, \qquad \boldsymbol{a}_j \in \{0,1\}^{4m+1} \quad (1 \leqslant j \leqslant 4),$$
$$t(\boldsymbol{a}_j) = m \quad (1 \leqslant j \leqslant 3), \qquad t(\boldsymbol{a}_4) = m+1, \qquad \boldsymbol{a}_1 + \cdots + \boldsymbol{a}_4 = \mathbf{1}_{4m+1}.$$

An elementary combinatorial argument in this instance therefore shows that $\mathcal{A}_{4m+1}$ is a (ν_{4m+1}, μ_{4m+1})-set of 3^m-decompositions of $\mathbf{1}_{4m+1}$, where
$$\nu_{4m+1} = \frac{(4m+1)!}{3!(m!)^3(m+1)!}$$
and
$$\mu_{4m+1} = \max\left\{\frac{(3m)!}{3!(m!)^3}, \frac{(3m+1)!}{2!(m!)^2(m+1)!}\right\} = \frac{(3m+1)!}{2(m!)^2(m+1)!}.$$

In the present situation, Lemmata 9.6 (iii) and 9.7 provide the bound
$$C(\boldsymbol{a}) \leqslant \frac{\widetilde{\tau}(\mathbf{1}_{4m+1})}{\lceil 2\nu_{4m+1}/\mu_{4m+1}\rceil} \cdot \frac{F(\boldsymbol{a}')}{3^m} \leqslant 2^{4m+1} 3^{1-m} \left[\frac{2}{3}\binom{4m+1}{m}\right]^{-1}. \tag{9.18}$$

When $m = 1$, the latter estimate yields $C(\boldsymbol{a}) \leqslant 2^5/\lceil 10/3 \rceil = 8$. For $m \geqslant 2$, meanwhile, the lower bound (9.17) shows that
$$\binom{4m+1}{m} = \binom{4m}{m}\frac{4m+1}{3m+1} \geqslant 4 \cdot 7^{m-1} \cdot \frac{9}{7} = 36 \cdot 7^{m-2}.$$

Consequently, we conclude from (9.18) that for $m \geqslant 2$, one has
$$C(\boldsymbol{a}) \leqslant 2^{4m} 3^{2-m} \binom{4m+1}{m}^{-1} \leqslant \frac{64}{9}\left(\frac{16}{21}\right)^{m-2} \leqslant \frac{64}{9}.$$

Thus we find that whenever $t(\boldsymbol{a}) = 4m+1$ with $m \geqslant 1$, then one has $C(\boldsymbol{a}) \leqslant 8$.

(vii) Suppose that $t(\boldsymbol{a}) = 4m+2$ with $m \geqslant 1$. We now write $\boldsymbol{a} = (\mathbf{1}_{4m+2}, \boldsymbol{a}')$ with $\boldsymbol{a}' \in V_2'$. Denote by $\mathcal{A}_{4m+2}$ the set of all the 3^m-decompositions $A = \{\boldsymbol{a}_1, \ldots, \boldsymbol{a}_4\}$ of $\mathbf{1}_{4m+2}$ such that $t(\boldsymbol{a}_j) = m$ for $j = 1$ and 2, and $t(\boldsymbol{a}_j) = m+1$ for $j = 3$ and 4. An elementary combinatorial argument confirms that $\mathcal{A}_{4m+2}$ is a (ν_{4m+2}, μ_{4m+2})-set of 3^m-decompositions of $\mathbf{1}_{4m+2}$, where
$$\nu_{4m+2} = \frac{(4m+2)!}{(2!m!(m+1)!)^2}$$

and
$$\mu_{4m+2} = \max\left\{\frac{(3m+1)!}{2!(m!)^2(m+1)!}, \frac{(3m+2)!}{2!m!((m+1)!)^2}\right\} = \frac{(3m+2)!}{2!m!((m+1)!)^2}.$$

We therefore deduce from Lemmata 9.6 (iii) and 9.7 that
$$C(\boldsymbol{a}) \leqslant \frac{\widetilde{\tau}(\boldsymbol{1}_{4m+2})}{2\nu_{4m+2}/\mu_{4m+2}} \cdot \frac{F(\boldsymbol{a}')}{3^m} \leqslant 2^{4m+2}3^{1-m}\binom{4m+2}{m}^{-1}.$$

For $m \geqslant 2$, the lower bound (9.17) implies that
$$\binom{4m+2}{m} = \binom{4m}{m}\frac{(4m+2)(4m+1)}{(3m+2)(3m+1)} \geqslant 4 \cdot 7^{m-1}\frac{10 \cdot 9}{8 \cdot 7} = 45 \cdot 7^{m-2},$$
whence
$$C(\boldsymbol{a}) \leqslant 2^{4m+2}3^{1-m}(45 \cdot 7^{m-2})^{-1} = \frac{2^{10}}{3 \cdot 45}\left(\frac{16}{21}\right)^{m-2} \leqslant \frac{1024}{135}.$$

Thus we conclude that whenever $t(\boldsymbol{a}) = 4m+2$ with $m \geqslant 2$, then one has $C(\boldsymbol{a}) < 8$.

When $m = 1$, or equivalently, when $t(\boldsymbol{a}) = 6$, we must be more explicit in our analysis. Observe now that $\mathcal{A}_6$ is a $(45, 15)$-set of 3-decompositions of $\boldsymbol{1}_6$. When $s(\boldsymbol{a}')$ is even, we deduce from Lemmata 9.6 (ii) and 9.7 that
$$C(\boldsymbol{a}) = C(\boldsymbol{1}_6, \boldsymbol{a}') \leqslant \frac{2^6}{2 \cdot 45/15} \cdot \frac{9/4}{3} = 8.$$

When $s(\boldsymbol{a}')$ is odd, on the other hand, we may write $\boldsymbol{a} = (\boldsymbol{1}_6, 2, \boldsymbol{a}'')$, where $\boldsymbol{a}'' \in V_2'$ and $s(\boldsymbol{a}'')$ is even. For each decomposition $A = (\boldsymbol{a}_1, \ldots, \boldsymbol{a}_4) \in \mathcal{A}_6$, one has $t(\boldsymbol{a}_1) = t(\boldsymbol{a}_2) = 1$ and $t(\boldsymbol{a}_3) = t(\boldsymbol{a}_4) = 2$, and so we find that the set
$$\{(\boldsymbol{a}_1, 1), (\boldsymbol{a}_2, 1), (\boldsymbol{a}_3, 0), (\boldsymbol{a}_4, 0)\}$$
provides a 9-decomposition of $(\boldsymbol{1}_6, 2)$. By taking the collection of all such sets for $A \in \mathcal{A}_6$, we obtain a $(45, 15)$-set of 9-decompositions of $(\boldsymbol{1}_6, 2)$. Consequently, by Lemmata 9.6 (ii) and 9.7, we obtain
$$C(\boldsymbol{a}) = C(\boldsymbol{1}_6, 2, \boldsymbol{a}'') \leqslant \frac{2^6 \cdot 3}{2 \cdot 45/15} \cdot \frac{9/4}{9} = 8.$$

We have therefore shown that $C(\boldsymbol{a}) \leqslant 8$ when $t(\boldsymbol{a}) = 6$, and this completes our treatment of case (vii).

(viii) Suppose that $t(\boldsymbol{a}) = 4m+3$ with $m \geqslant 1$. In this case we apply an argument closely resembling that of case (vii). Put $\boldsymbol{a} = (\boldsymbol{1}_{4m+3}, \boldsymbol{a}')$ with $\boldsymbol{a}' \in V_2'$, and denote by $\mathcal{A}_{4m+3}$ the set of all the 3^m-decompositions $A = \{\boldsymbol{a}_1, \ldots, \boldsymbol{a}_4\}$ of $\boldsymbol{1}_{4m+3}$ such that $t(\boldsymbol{a}_1) = m$ and $t(\boldsymbol{a}_j) = m+1$ for $2 \leqslant j \leqslant 4$. An elementary combinatorial argument again confirms that $\mathcal{A}_{4m+3}$ is a (ν_{4m+3}, μ_{4m+3})-set of 3^m-decompositions of $\boldsymbol{1}_{4m+3}$, where
$$\nu_{4m+3} = \frac{(4m+3)!}{3!m!((m+1)!)^3}$$

and
$$\mu_{4m+3} = \max\left\{\frac{(3m+2)!}{2!m!((m+1)!)^2}, \frac{(3m+3)!}{3!((m+1)!)^3}\right\} = \frac{(3m+3)!}{3!((m+1)!)^3}.$$

We therefore deduce from Lemmata 9.6 (iii) and 9.7 that

$$C(\boldsymbol{a}) \leqslant \frac{\widetilde{\tau}(\boldsymbol{1}_{4m+3})}{2\nu_{4m+3}/\mu_{4m+3}} \cdot \frac{F(\boldsymbol{a}')}{3^m} \leqslant 2^{4m+2}3^{1-m}\binom{4m+3}{m}^{-1}. \qquad (9.19)$$

The inequality (9.17) now implies that

$$\binom{4m+3}{m} = \frac{1}{4}\binom{4(m+1)}{m+1} \geqslant 7^m,$$

and so we find from (9.19) that for $m \geqslant 2$, one has

$$C(\boldsymbol{a}) \leqslant 12\left(\frac{16}{21}\right)^m \leqslant \frac{1024}{147}.$$

When $m = 1$ we have $t(\boldsymbol{a}) = 7$, and must again proceed more carefully. In this case $\mathcal{A}_7$ is a $(105, 15)$-set of 3-decompositions of $\boldsymbol{1}_7$. When $s(\boldsymbol{a}')$ is even, it follows from Lemmata 9.6 (ii) and 9.7 that

$$C(\boldsymbol{a}) = C(\boldsymbol{1}_7, \boldsymbol{a}') \leqslant \frac{2^7}{2 \cdot 105/15} \cdot \frac{9/4}{3} = \frac{48}{7}.$$

When $s(\boldsymbol{a}')$ is odd, meanwhile, we write $\boldsymbol{a} = (\boldsymbol{1}_7, 2, \boldsymbol{a}'')$, where $\boldsymbol{a}'' \in V_2'$ and $s(\boldsymbol{a}'')$ is even. For each decomposition $A = (\boldsymbol{a}_1, \ldots, \boldsymbol{a}_4) \in \mathcal{A}_7$, one has $t(\boldsymbol{a}_1) = 1$ and $t(\boldsymbol{a}_j) = 2$ $(2 \leqslant j \leqslant 4)$, and so we find that the set

$$\{(\boldsymbol{a}_1, 2), (\boldsymbol{a}_2, 0), (\boldsymbol{a}_3, 0), (\boldsymbol{a}_4, 0)\}$$

provides a 9-decomposition of $(\boldsymbol{1}_7, 2)$. Collecting such decompositions for $A \in \mathcal{A}_7$, we obtain a $(105, 15)$-set of 9-decompositions of $(\boldsymbol{1}_7, 2)$. Thus, by Lemmata 9.6 (ii) and 9.7, we may conclude that

$$C(\boldsymbol{a}) = C(\boldsymbol{1}_7, 2, \boldsymbol{a}'') \leqslant \frac{2^7 \cdot 3}{2 \cdot 105/15} \cdot \frac{9/4}{9} = \frac{48}{7}.$$

We have therefore shown that $C(\boldsymbol{a}) < 8$ when $t(\boldsymbol{a}) = 7$, and this completes our analysis of case (viii).

We conclude by noting that the discussions of cases (i) through (viii) above demonstrate that $C(\boldsymbol{a}) \leqslant 8$ whenever $\boldsymbol{a} \in V_1$, and so this completes the proof of the lemma. □

This completes the third phase of our argument, so that in view of the discussion in the preamble to Lemma 9.7, and that in the paragraph containing (9.5), the desired conclusion (9.1) is now available with $C = 8$. We summarise the deliberations of this section in the form of a lemma.

LEMMA 9.10. — *Denote the number of divisors of n by $\tau(n)$, and let g be the multiplicative function defined by means of* (9.2). *Then for every natural number n one has*
$$\tau(n) \leqslant 8 \sum_{\substack{d \mid n \\ d \leqslant n^{1/4}}} g(d).$$

CHAPTER 10

AN INVESTIGATION OF CERTAIN CONGRUENCES

Our next goal is to establish the mean value estimates recorded in Lemmata 2.4 and 2.5. Before embarking on this mission in §11, we require some estimates associated with the number of solutions of certain congruences, and these we prepare in the current section. Our first lemma concerns the number $\rho(d)$ of solutions of the congruence

$$x_1^4 + x_2^4 \equiv x_3^4 + x_4^4 \pmod{d}, \tag{10.1}$$

with $1 \leqslant x_j \leqslant d$ ($1 \leqslant j \leqslant 4$).

LEMMA 10.1. — *The function $\rho(d)$ satisfies the following properties.*

(i) *One has*

$$\rho(2^v) = \begin{cases} 8, & \text{when } v = 1, \\ 3 \cdot 2^{4v-3}, & \text{when } 2 \leqslant v \leqslant 4. \end{cases}$$

Further, when $u \geqslant 0$ and $1 \leqslant v \leqslant 4$, one has

$$\rho(2^{4u+v}) = 5u \cdot 2^{12u+3v} + 2^{12u}\rho(2^v).$$

(ii) *One has $\rho(3) = 33$ and $\rho(5) = 321$. Also, defining $\kappa(p)$ and b_p as in (5.5) and (5.15) respectively, for each odd prime p one has the relation*

$$p^3 \leqslant \rho(p) \leqslant p^3 + b_p \kappa(p)^2 p^2 (p-1).$$

Finally, when p is an odd prime, $u \geqslant 0$ and $1 \leqslant v \leqslant 4$, one has

$$\rho(p^{4u+v}) = (u+1)p^{3(4u+v-1)}(\rho(p) - 1) + p^{12u+4v-4}.$$

Proof. — We begin by disposing of the simplest cases of the lemma with minimal effort. Of course, one instantaneously verifies that $\rho(2) = 8$. Next, we define $\sigma_q(m)$ to be the number of solutions of the congruence $x^4 + y^4 \equiv m \pmod{q}$ with $1 \leqslant x, y \leqslant q$, and we observe that

$$\rho(q) = \sum_{m=1}^{q} \sigma_q(m)^2.$$

Suppose for the moment that q is one of 3, 4, 5, 8 or 16. Then one has $x^4 \equiv 1 \pmod{q}$ whenever $(x, q) = 1$, and otherwise $x^4 \equiv 0 \pmod{q}$. We therefore find that $\sigma_q(m)$ is zero unless m is one of 0, 1 or 2, and thus an elementary computation reveals that

$$\rho(p) = 1^2 + \binom{2}{1}^2 (p-1)^2 + (p-1)^4 \qquad (p = 3, 5),$$

$$\rho(2^v) = 2^{4v-4} + \binom{2}{1}^2 2^{4v-4} + 2^{4v-4} \qquad (v = 2, 3, 4).$$

The initial conclusions of parts (i) and (ii) of the lemma now follow immediately.

Suppose next that p is an odd prime, and recall the definition (5.1) of the exponential sum $S(q, a)$. By orthogonality, one has

$$\rho(p) = p^{-1} \sum_{a=1}^{p} |S(p,a)|^4 = p^3 + p^{-1} \sum_{a=1}^{p-1} |S(p,a)|^4.$$

On recalling Lemmata 5.1 and 5.3, we find that

$$0 \leqslant p^{-1} \sum_{a=1}^{p-1} |S(p,a)|^4 \leqslant p\kappa(p)^2 \sum_{a=1}^{p-1} |S(p,a)|^2 = b_p \kappa(p)^2 p^2 (p-1),$$

whence the second conclusion of part (ii) of the lemma follows.

We next turn our attention to the final conclusions of parts (i) and (ii) of the lemma. When d is a natural number, denote by $\rho^*(d)$ the number of solutions $\boldsymbol{x}$ of the congruence (10.1) counted by $\rho(d)$ with $(x_j, d) = 1$ for some j. Then defining $\gamma = \gamma(p)$ as in the statement of Lemma 6.1, we find from the latter lemma that when $l \geqslant \gamma$,

$$\rho^*(p^l) = p^{3(l-\gamma)} \rho^*(p^\gamma). \tag{10.2}$$

We therefore deduce from the definitions of $\rho(d)$ and $\rho^*(d)$ that for $l \geqslant 4$,

$$\rho(p^l) = \rho^*(p^l) + (p^3)^4 \rho(p^{l-4}) = p^{3(l-\gamma)} \rho^*(p^\gamma) + p^{12} \rho(p^{l-4}). \tag{10.3}$$

A u-fold application of this formula demonstrates that for $u \geqslant 0$ and $1 \leqslant v \leqslant 4$, one has

$$\rho(p^{4u+v}) = u p^{3(4u+v-\gamma)} \rho^*(p^\gamma) + p^{12u} \rho(p^v). \tag{10.4}$$

In the case $p = 2$, we see that $\rho(2^4) - \rho^*(2^4)$ is equal to the number of quadruples $(x_1, \ldots, x_4)$ with each x_j even and satisfying $1 \leqslant x_j \leqslant 16$. Thus it follows from (10.4) that when $u \geqslant 0$ and $1 \leqslant v \leqslant 4$,

$$\rho(2^{4u+v}) = u \cdot 2^{3(4u+v-4)} (\rho(2^4) - (2^3)^4) + 2^{12u} \rho(2^v),$$

whence the final conclusion of part (i) of the lemma follows from the first conclusion of that part. When p is odd, meanwhile, we deduce from (10.2) that when $1 \leqslant v \leqslant 4$,

$$\rho(p^v) = \rho^*(p^v) + p^{4(v-1)} = p^{3(v-1)} \rho^*(p) + p^{4(v-1)}. \tag{10.5}$$

MÉMOIRES DE LA SMF 100

An application of the relation (10.4) yields the conclusion that for $u \geqslant 0$ and $1 \leqslant v \leqslant 4$, one has
$$\rho(p^{4u+v}) = (u+1)p^{3(4u+v-1)}\rho^*(p) + p^{12u+4v-4},$$
and the final conclusion of part (ii) of the lemma follows from the observation that $\rho^*(p) = \rho(p) - 1$. □

We next establish a variant of the previous lemma relevant to the number, $r(d) = r(d;\varepsilon)$, of solutions of the congruence
$$\frac{1}{2}\big((2x_1+\varepsilon)^4 + (2x_2+\varepsilon)^4 - (2x_3+\varepsilon)^4 - (2x_4+\varepsilon)^4\big) \equiv 0 \pmod{d}, \qquad (10.6)$$
with $1 \leqslant x_j \leqslant d$ ($1 \leqslant j \leqslant 4$).

LEMMA 10.2. — *Suppose that $\varepsilon \in \{0,1\}$. Then the following conclusions hold.*
  (i) *When d is odd, one has $r(d) = \rho(d)$.*
  (ii) *When $1 \leqslant l \leqslant 3$, one has $r(2^l) = 2^{4l}$, and for $u \geqslant 0$ and $1 \leqslant v \leqslant 4$,*
$$r(2^{4u+v+3}) \leqslant 5u \cdot 2^{3(4u+v+4)} + 2^{12(u+1)}\rho(2^v).$$

Proof. — We begin by noting that when d is odd, a change of variables confirms that $r(d) = \rho(d)$, thereby establishing part (i) of the lemma. Observe next that when $d = 2^l$ with $l \leqslant 3$, it follows from (5.11) that the congruence (10.6) is automatically satisfied for every choice of $\boldsymbol{x}$. In particular, one has $r(2^l) = 2^{4l}$ for $1 \leqslant l \leqslant 3$. Suppose then that $l \geqslant 4$. When $\varepsilon = 0$, we see at once that $r(2^l;0) = 2^{12}\rho(2^{l-3})$. On writing $l - 3 = 4u + v$ with $u \geqslant 0$ and $1 \leqslant v \leqslant 4$, we therefore deduce from Lemma 10.1 (i) that
$$r(2^{4u+v+3};0) = 2^{12}\big(5u \cdot 2^{12u+3v} + 2^{12u}\rho(2^v)\big),$$
and so the desired conclusion follows for $\varepsilon = 0$. When $\varepsilon = 1$, in the meantime, we may observe that $r(2^l;1)$ is equal to $r'(2^{l+1})$, where we take $r'(2^m)$ to be the number of solutions of the congruence
$$y_1^4 + y_2^4 \equiv y_3^4 + y_4^4 \pmod{2^m},$$
with $1 \leqslant y_j \leqslant 2^m$ and y_j odd for $1 \leqslant j \leqslant 4$. It is a consequence of Lemma 6.1 that $r'(2^m) = 2^{3(m-4)}r'(2^4)$ for $m \geqslant 4$, so that the trivial conclusion $r'(2^4) = (2^3)^4$ leads to the relation
$$r(2^l;1) = r'(2^{l+1}) = 2^{3(l-3)+12} \qquad (l \geqslant 4).$$
On writing $l - 3 = 4u + v$ as above, and noting that Lemma 10.1 (i) provides the lower bound $\rho(2^v) \geqslant 2^{3v}$ for $1 \leqslant v \leqslant 4$, we arrive at the upper bound
$$r(2^{4u+v+3};1) = 2^{3(4u+v)+12} \leqslant 2^{12(u+1)}\rho(2^v).$$
The desired conclusion therefore follows also for $\varepsilon = 1$, and this completes the proof of the lemma. □

Finally, we require an analogue of Lemma 10.1 in which the congruence (10.1) is replaced by a new congruence stemming from the use of the identity (1.5). In this context, let $s(d) = s(d; \zeta)$ denote the number of solutions of the congruence

$$\frac{1}{4}\left(30(2x_1 + \zeta)^4 - 30(2x_2 + \zeta)^4 + (2y_1 + 1)^4 - (2y_2 + 1)^4\right) \equiv 0 \pmod{d},$$

with $1 \leq x_j, y_j \leq d$ ($j = 1, 2$).

LEMMA 10.3. — *Suppose that $\zeta \in \{0, 1\}$. Then the following conclusions hold.*

(i) *When d is a natural number with $(d, 30) = 1$, one has $d^3 \leq s(d) \leq \rho(d)$.*

(ii) *When $p = 3$ or 5, one has $s(p) = p^2(p^2 - 2p + 2)$, and when $l \geq 1$ one has $s(p^l) < \frac{3}{2}\rho(p^l)$.*

(iii) *One has*

$$s(2^l) = \begin{cases} 2^{4l}, & \text{when } l = 1, 2, \\ 2^{3l+2}, & \text{when } l \geq 3. \end{cases}$$

Proof. — We observe first that when d is odd, a change of variables demonstrates that $s(d)$ is equal to the number of solutions of the congruence

$$30(x_1^4 - x_2^4) \equiv y_1^4 - y_2^4 \pmod{d}, \tag{10.7}$$

with $1 \leq x_j, y_j \leq d$ ($j = 1, 2$). On recalling (5.1), we thus deduce by orthogonality that

$$s(d) = d^{-1} \sum_{a=1}^{d} |S(d, 30a) S(d, a)|^2.$$

On the one hand, the contribution of the term $a = d$ in this last sum suffices to confirm the lower bound $s(d) \geq d^3$. On the other hand, since orthogonality provides the relation

$$\rho(d) = d^{-1} \sum_{a=1}^{d} |S(d, a)|^4,$$

it follows from Cauchy's inequality together with a change of variable that when $(d, 30) = 1$, one has

$$s(d) \leq \left(d^{-1} \sum_{a=1}^{d} |S(d, 30a)|^4\right)^{1/2} \left(d^{-1} \sum_{a=1}^{d} |S(d, a)|^4\right)^{1/2}$$

$$= d^{-1} \sum_{a=1}^{d} |S(d, a)|^4 = \rho(d).$$

This completes the proof of part (i) of the lemma.

We next turn to part (ii) of the lemma, and suppose for the moment that $p = 3$ or 5. For $j = 1, 2, 3$ and $l \geq 0$, let $s_j(p^l)$ denote the number of solutions of the congruence

(10.7) with $d = p^l$, $1 \leqslant x_i$, $y_i \leqslant p^l$ ($i = 1, 2$), subject to the list of conditions C_j, where C_j is given as follows:

$$C_1 \: : \: p \nmid y_1 \text{ or } p \nmid y_2,$$
$$C_2 \: : \: p|y_1, \, p|y_2 \text{ and } p \nmid x_1 \text{ or } p \nmid x_2,$$
$$C_3 \: : \: p|x_1, \, p|x_2, \, p|y_1, \, p|y_2.$$

Then we have
$$s(p^l) = s_1(p^l) + s_2(p^l) + s_3(p^l). \tag{10.8}$$

Further, one has the instant relation
$$s_3(p^l) = \begin{cases} p^{4(l-1)}, & \text{when } 1 \leqslant l \leqslant 4, \\ p^{12} s(p^{l-4}), & \text{when } l \geqslant 5. \end{cases} \tag{10.9}$$

We estimate $s_1(p^l)$ and $s_2(p^l)$ as follows. Observing that when $d = p$, the congruence (10.7) simplifies to $y_1^4 \equiv y_2^4 \pmod{p}$, we find that $s_1(p) = p^2(p-1)^2$. Thus, appealing to Lemma 6.1, we deduce that when $l \geqslant 1$ one has

$$s_1(p^l) = p^{3(l-1)} s_1(p) = p^{3l-1}(p-1)^2. \tag{10.10}$$

Next, when $l \geqslant 0$, we denote by $s_2'(p^l)$ the number of solutions of the congruence
$$30 p^{-1}(x_1^4 - x_2^4) \equiv p^3 (y_1^4 - y_2^4) \pmod{p^l},$$
with $1 \leqslant x_i, y_i \leqslant p^l$ and $p \nmid x_i$ ($i = 1, 2$). Plainly, one has $s_2'(p) = p^2(p-1)^2$, and so it follows from Lemma 6.1 that when $l \geqslant 1$, one has
$$s_2'(p^l) = p^{3l-1}(p-1)^2.$$

We therefore deduce that when $l \geqslant 2$,
$$s_2(p^l) = p^2 s_2'(p^{l-1}) = p^{3l-2}(p-1)^2, \tag{10.11}$$
while the relation
$$s_2(p) = p^2 - 1 \tag{10.12}$$
is immediate from the definition of $s_2(p)$.

On collecting together (10.8)-(10.12), we obtain
$$s(p) = p^2(p-1)^2 + (p^2 - 1) + 1 = p^2(p^2 - 2p + 2), \tag{10.13}$$
which confirms the first claim of part (ii) of the lemma, and one also obtains the relations
$$s(p^l) = p^{3l-2}(p+1)(p-1)^2 + \begin{cases} p^{4(l-1)}, & \text{when } 2 \leqslant l \leqslant 4, \\ p^{12} s(p^{l-4}), & \text{when } l \geqslant 5. \end{cases} \tag{10.14}$$

In order to establish the remaining conclusion of part (ii) of the lemma, we note initially that by Lemma 10.1 (ii) and (10.13), one has $s(p) < \frac{3}{2}\rho(p)$ for $p = 3, 5$.

Also, recalling the definition of $\rho^*(d)$ from the proof of Lemma 10.1, and noting that $\rho^*(p) = \rho(p) - 1$, one finds from Lemma 10.1 (ii) that for $p = 3$ and 5 one has

$$p(p+1)(p-1)^2 = \frac{3}{2}\rho^*(p). \tag{10.15}$$

On recalling (10.5), we now conclude from (10.14) and (10.15) that for $2 \leqslant l \leqslant 4$, one has

$$s(p^l) = \frac{3}{2} p^{3(l-1)} \rho^*(p) + p^{4(l-1)} < \frac{3}{2} \rho(p^l).$$

When $l \geqslant 5$, meanwhile, we may make use of (10.3) with $\gamma = 1$ in combination with (10.14) and (10.15) to deduce that whenever $s(p^{l-4}) < \frac{3}{2}\rho(p^{l-4})$, then one has

$$s(p^l) = \frac{3}{2} p^{3(l-1)} \rho^*(p) + p^{12} s(p^{l-4})$$
$$< \frac{3}{2}\left(p^{3(l-1)}\rho^*(p) + p^{12}\rho(p^{l-4})\right) = \frac{3}{2}\rho(p^l).$$

The final conclusion of part (ii) of the lemma consequently follows by induction on l, with our previous conclusions providing the basis of the induction.

It now remains only to establish part (iii) of the lemma. The desired conclusion for $l = 1$ and 2 is immediate from (5.11). In order to handle the cases with $l \geqslant 3$, we introduce the function $s'(2^m)$, which we define to be the number of solutions of the congruence

$$30x_1^4 - 30x_2^4 + y_1^4 - y_2^4 \equiv 0 \pmod{2^m},$$

with $1 \leqslant x_j, y_j \leqslant 2^m$, $x_j \equiv \zeta \pmod{2}$ and $y_j \equiv 1 \pmod{2}$ ($j = 1, 2$). A moment's reflection yields the relation $s(2^l) = 2^{-4} s'(2^{l+2})$. Moreover, the solutions $\boldsymbol{x}$, $\boldsymbol{y}$ counted by $s'(2^m)$ satisfy the property that y_1 is odd, whence by Lemma 6.1 one has

$$s'(2^m) = 2^{3(m-4)} s'(2^4) \qquad (m \geqslant 4).$$

In view of (5.11), we have $s'(2^4) = (2^3)^4$, and thus we deduce that whenever $l \geqslant 3$, one has

$$s(2^l) = 2^{-4} s'(2^{l+2}) = 2^{-4+3(l-2)+12} = 2^{3l+2}.$$

This completes the proof of part (iii) of the lemma, and completes our discussion of Lemma 10.3. □

The final three lemmata of this section provide weighted sums of the functions $r(d)$ and $s(d)$ occurring in Lemmata 10.2 and 10.3, together with our surrogate for the divisor function, $g(d)$, that was central to the discussion of the previous section. We begin with a technical lemma that simplifies our subsequent detailed investigations specific to the functions $r(d)$ and $s(d)$.

LEMMA 10.4. — *Let $\lambda(d)$ be a multiplicative function satisfying the condition that $\lambda(d) \geqslant 0$ for all natural numbers d, and also satisfying the property that for every* prime p, one has $\lambda(p) \geqslant p^3$. Suppose that X is a real number with $X \geqslant 1$, and that k is a real number with $0 \leqslant k \leqslant 3$. Then one has

$$\sum_{1 \leqslant d \leqslant X} \frac{g(d)\lambda(d)}{d^k} \leqslant X^{4-k} \prod_{p \leqslant X} \left(1 - \frac{1}{p} + \sum_{l=1}^{\operatorname{Log}_p X} \frac{g(p^l)\lambda(p^l)}{p^{4l}} \right),$$

where we write $\operatorname{Log}_p X$ for $[(\log X)/(\log p)]$.

Proof. — Let $\mathcal{D}$ denote the set of squarefull numbers, by which we mean the set of natural numbers n with the property that whenever p is a prime number with $p|n$, then necessarily $p^2|n$. Every natural number d may be written uniquely in the shape $d = d_1 d_2$, where d_1 is squarefree, d_2 is squarefull, and $(d_1, d_2) = 1$. On writing also $\mathcal{S}$ for the set of squarefree numbers, we see that

$$\sum_{d \leqslant X} \frac{g(d)\lambda(d)}{d^k} = \sum_{\substack{d_2 \leqslant X \\ d_2 \in \mathcal{D}}} \frac{g(d_2)\lambda(d_2)}{d_2^k} \sum_{\substack{d_1 \leqslant X/d_2 \\ (d_1,d_2)=1 \\ d_1 \in \mathcal{S}}} \frac{g(d_1)\lambda(d_1)}{d_1^k}. \tag{10.16}$$

We analyse the inner sum of the right hand side of (10.16) first. When q is squarefree, define the function $\theta(q)$ by

$$\theta(q) = \prod_{p|q} (\lambda(p)p^{-3} - 1).$$

Note that in view of the hypothesis on $\lambda(p)$ imposed in the statement of the lemma, one has $\theta(q) \geqslant 0$ for every q. Plainly, for each prime p one has $\lambda(p) = p^3(1 + \theta(p))$, and hence we deduce from the multiplicative property of $\lambda(d)$ that for squarefree d_1, one has

$$\lambda(d_1) = d_1^3 \prod_{p|d_1} (1 + \theta(p)) = d_1^3 \sum_{qh=d_1} \theta(q).$$

Consequently, the innermost sum on the right hand side of (10.16) may be written in the form

$$\sum_{\substack{d_1 \leqslant X/d_2 \\ (d_1,d_2)=1 \\ d_1 \in \mathcal{S}}} \frac{g(d_1)\lambda(d_1)}{d_1^k} = \sum_{\substack{q \leqslant X/d_2 \\ (q,d_2)=1 \\ q \in \mathcal{S}}} g(q)\theta(q) q^{3-k} \sum_{\substack{h \leqslant X/(qd_2) \\ (h,qd_2)=1 \\ h \in \mathcal{S}}} g(h) h^{3-k}. \tag{10.17}$$

We next tackle the innermost sum on the right hand side of (10.17). Since $g(p) = 3 = 1 + \tau(p)$, where $\tau(n)$ again denotes the divisor function, we have for each squarefree number h the relation

$$g(h) = \prod_{p|h}(1 + \tau(p)) = \sum_{mn=h} \tau(m).$$

Thus we deduce that

$$\sum_{\substack{h \leqslant X/(qd_2) \\ (h,qd_2)=1 \\ h \in \mathcal{S}}} g(h) h^{3-k} = \sum_{\substack{m \leqslant X/(qd_2) \\ (m,qd_2)=1 \\ m \in \mathcal{S}}} \tau(m) m^{3-k} \sum_{\substack{n \leqslant X/(qmd_2) \\ (n,qmd_2)=1 \\ n \in \mathcal{S}}} n^{3-k}.$$

But for $0 \leqslant k \leqslant 3$, the innermost sum in the last expression is plainly at most $(X/(qmd_2))^{4-k}$, and hence one obtains the estimate

$$\sum_{\substack{h \leqslant X/(qd_2) \\ (h,qd_2)=1 \\ h \in \mathcal{S}}} g(h) h^{3-k} \leqslant \left(\frac{X}{d_2 q}\right)^{4-k} \sum_{\substack{m \leqslant X/(qd_2) \\ (m,qd_2)=1 \\ m \in \mathcal{S}}} \frac{\tau(m)}{m}$$

$$\leqslant \left(\frac{X}{d_2 q}\right)^{4-k} \prod_{\substack{p \leqslant X/d_2 \\ p \nmid qd_2}} (1+2/p). \qquad (10.18)$$

We now substitute from (10.18) into (10.17) to deduce that

$$\sum_{\substack{d_1 \leqslant X/d_2 \\ (d_1,d_2)=1 \\ d_1 \in \mathcal{S}}} \frac{g(d_1)\lambda(d_1)}{d_1^k}$$

$$\leqslant \left(\frac{X}{d_2}\right)^{4-k} \prod_{\substack{p \leqslant X/d_2 \\ p \nmid d_2}} \left(1+\frac{2}{p}\right) \sum_{\substack{q \leqslant X/d_2 \\ (q,d_2)=1 \\ q \in \mathcal{S}}} \frac{g(q)\theta(q)}{q} \prod_{p|q} \left(1+\frac{2}{p}\right)^{-1}$$

$$\leqslant \left(\frac{X}{d_2}\right)^{4-k} \prod_{\substack{p \leqslant X/d_2 \\ p \nmid d_2}} \left(1+\frac{2}{p}\right) \prod_{\substack{\varpi \leqslant X/d_2 \\ \varpi \nmid d_2}} \left(1 + \frac{g(\varpi)\theta(\varpi)}{\varpi}\left(1+\frac{2}{\varpi}\right)^{-1}\right),$$

where ϖ implicitly denotes a prime number. On writing

$$\theta_1(p) = 1 + \frac{2}{p} + \frac{g(p)\theta(p)}{p} = 1 - \frac{1}{p} + \frac{g(p)\lambda(p)}{p^4}, \qquad (10.19)$$

we may conclude that

$$\sum_{\substack{d_1 \leqslant X/d_2 \\ (d_1,d_2)=1 \\ d_1 \in \mathcal{S}}} \frac{g(d_1)\lambda(d_1)}{d_1^k} \leqslant \left(\frac{X}{d_2}\right)^{4-k} \prod_{\substack{p \leqslant X/d_2 \\ p \nmid d_2}} \theta_1(p). \qquad (10.20)$$

Finally, we substitute from (10.20) into (10.16) to obtain

$$\sum_{d \leqslant X} \frac{g(d)\lambda(d)}{d^k} \leqslant X^{4-k} \left(\prod_{p \leqslant X} \theta_1(p)\right) \sum_{\substack{d_2 \leqslant X \\ d_2 \in \mathcal{D}}} \frac{g(d_2)\lambda(d_2)}{d_2^4} \prod_{\varpi | d_2} \theta_1(\varpi)^{-1}$$

$$\leqslant X^{4-k} \left(\prod_{p \leqslant X} \theta_1(p)\right) \prod_{\varpi \leqslant X} \left(1 + \sum_{l=2}^{\mathrm{Log}_\varpi X} \frac{g(\varpi^l)\lambda(\varpi^l)}{\varpi^{4l}} \theta_1(\varpi)^{-1}\right)$$

$$= X^{4-k} \prod_{p \leqslant X} \left(\theta_1(p) + \sum_{l=2}^{\mathrm{Log}_p X} \frac{g(p^l)\lambda(p^l)}{p^{4l}}\right).$$

The desired conclusion now follows immediately from (10.19). □

The moment has arrived to extract the estimates critical to our proofs, in the next section, of Lemmata 2.4 and 2.5. We begin with a lemma that provides the key ingredient in our proof of the former lemma.

LEMMA 10.5. — *Suppose that* $X \geqslant 10^{25}$ *and* $\varepsilon \in \{0,1\}$. *Then*
$$\sum_{d \leqslant X} g(d) r(d;\varepsilon) d^{-4} < 125 (\log X)^3,$$
and, whenever $0 \leqslant k \leqslant 3$, *one has also*
$$\sum_{d \leqslant X} g(d) r(d;\varepsilon) d^{-k} < 256 X^{4-k} (\log X)^2.$$

Proof. — Let $\varepsilon \in \{0,1\}$, and write $r(d) = r(d;\varepsilon)$. Putting
$$H(p) = \sum_{l=1}^{\infty} g(p^l) r(p^l) p^{-4l}, \qquad (10.21)$$
it follows from the multiplicative properties of $g(d)$ and $r(d)$ that
$$\sum_{d \leqslant X} g(d) r(d) d^{-4} \leqslant \prod_{p \leqslant X} (1 + H(p)). \qquad (10.22)$$

We explicitly compute upper bounds for the factors $1 + H(p)$, for each prime p, and from these bounds the first conclusion of the lemma will follow directly.

We begin with the even prime. By the definition (9.2) of $g(d)$, we have
$$1 + H(2) = 1 + 3 + 3 + 9 + 9 \sum_{u=0}^{\infty} \sum_{v=1}^{4} \frac{r(2^{4u+v+3})}{2^{4(4u+v+3)}} + 18 \sum_{l \in \{7,9,11\}} \frac{r(2^l)}{2^{4l}}.$$

Thus, on making use of Lemmata 10.1 (i) and 10.2 (ii), we find that
$$1 + H(2) \leqslant 16 + 18 \cdot \frac{133}{256} + 9 \sum_{u=0}^{\infty} \sum_{v=1}^{4} \left(5u \cdot 2^{-4u-v} + 2^{-4u-4v} \rho(2^v) \right). \qquad (10.23)$$

Using the formulae
$$\sum_{m=0}^{\infty} z^m = \frac{1}{1-z} \quad \text{and} \quad \sum_{m=1}^{\infty} m z^{m-1} = \frac{1}{(1-z)^2}, \qquad (10.24)$$
which are valid for $|z| < 1$, one finds that
$$\sum_{u=0}^{\infty} u 2^{-4u} = \frac{16}{225}, \qquad \sum_{u=0}^{\infty} 2^{-4u} = \frac{16}{15}.$$

Also, we have
$$\sum_{v=1}^{4} 2^{-v} = \frac{15}{16},$$

and on making use of Lemma 10.1 once again, we obtain

$$\sum_{v=1}^{4} 2^{-4v}\rho(2^v) = 2^{-4} \cdot 2^3 + \sum_{v=2}^{4} 3 \cdot 2^{-3} = \frac{13}{8}.$$

Consequently, the upper bound (10.23) becomes

$$1 + H(2) \leqslant 16 + \frac{1197}{128} + 9\left(5 \cdot \frac{16}{225} \cdot \frac{15}{16} + \frac{16}{15} \cdot \frac{13}{8}\right)$$
$$= \frac{28129}{640}. \tag{10.25}$$

We next turn our attention to the odd primes p. On making use of Lemmata 10.1 (ii) and 10.2 (i) together with the upper bound $g(p^l) \leqslant 27$, valid for $l \geqslant 9$, we obtain

$$\sum_{l=9}^{\infty} \frac{g(p^l)r(p^l)}{p^{4l}}$$
$$\leqslant 27 \sum_{u=2}^{\infty} \sum_{v=1}^{4} p^{-4(4u+v)} \left((u+1)p^{12u+3v-3}(\rho(p)-1) + p^{12u+4v-4}\right),$$

whence by (10.24) we deduce that

$$\sum_{l=9}^{\infty} \frac{g(p^l)r(p^l)}{p^{4l}} \leqslant 27 \left(\frac{\rho(p)-1}{p^3}\right)\left(\frac{3-2p^{-4}}{p^4(p^4-1)(p-1)}\right) + \frac{108}{p^8(p^4-1)}. \tag{10.26}$$

Furthermore, on making use of the definition of $g(p^l)$ together with Lemmata 10.1 (ii) and 10.2 (i), we obtain

$$\sum_{l=1}^{8} \frac{g(p^l)r(p^l)}{p^{4l}} = \left(\frac{\rho(p)-1}{p^3}\right)\left(\frac{3}{p} + \frac{3}{p^2} + \frac{9}{p^3} + \frac{9}{p^4} + \frac{18}{p^5} + \frac{18}{p^6} + \frac{54}{p^7} + \frac{18}{p^8}\right)$$
$$+ \frac{24}{p^4} + \frac{54}{p^8}. \tag{10.27}$$

On substituting (10.26) and (10.27) into (10.21), and making use of the explicit values for $\rho(3)$ and $\rho(5)$ provided in Lemma 10.1 (ii), we therefore deduce that

$$H(3) < 2.57 \quad \text{and} \quad H(5) < 2.123. \tag{10.28}$$

For primes p of intermediate size, we note that Lemma 10.1 (ii) supplies the bound

$$\frac{\rho(p)-1}{p^3} < 1 + b_p \kappa(p)^2.$$

On combining (10.21), (10.26) and (10.27), we thus obtain for $p \geqslant 7$ the upper bound

$$H(p) < (1 + b_p \kappa(p)^2)(3p^{-1} + 3p^{-2} + 10.73p^{-3}) + 24.03p^{-4}. \tag{10.29}$$

Applying this bound in combination with (5.5) and (5.15), a direct computation reveals that
$$\prod_{\substack{7\leqslant p\leqslant 79 \\ p\equiv 3\ (\text{mod}\ 4)}} (1+H(p)) < 4.35, \tag{10.30}$$
and
$$\prod_{\substack{13\leqslant p\leqslant 73 \\ p\equiv 1\ (\text{mod}\ 4)}} (1+H(p)) < 3.88. \tag{10.31}$$

When p is a prime with $p \geqslant 83$, it follows from (5.5) and (5.15) that $b_p\kappa(p)^2 = b_p^3/p$, and thus we derive from (10.29) the upper bound
$$1 + H(p) < 1 + 3p^{-1} + (3.038b_p^3 + 3.133)p^{-2}$$
$$< \exp\bigl(3p^{-1} + (3.038b_p^3 + 3.133)p^{-2}\bigr). \tag{10.32}$$

Thus we have
$$\prod_{83\leqslant p\leqslant X} (1+H(p)) < \exp\left(3\sum_{83\leqslant p\leqslant X}\frac{1}{p} + 6.171\sum_{\substack{p\geqslant 83 \\ p\equiv 3\ (\text{mod}\ 4)}}\frac{1}{p^2} + 85.16\sum_{\substack{p\geqslant 89 \\ p\equiv 1\ (\text{mod}\ 4)}}\frac{1}{p^2}\right). \tag{10.33}$$

But as a trivial consequence of Lemma 5.4, one has
$$\sum_{\substack{p\geqslant 83 \\ p\equiv 3\ (\text{mod}\ 4)}} \frac{1}{p^2} < \frac{1}{4\cdot 79} = \frac{1}{316} \quad\text{and}\quad \sum_{\substack{p\geqslant 89 \\ p\equiv 1\ (\text{mod}\ 4)}} \frac{1}{p^2} < \frac{1}{4\cdot 85} = \frac{1}{340}. \tag{10.34}$$

Also, on applying Lemma 3.2 together with a direct computation, one finds that for $X \geqslant 10^{25}$, one has
$$\sum_{83\leqslant p\leqslant X}\frac{1}{p} < \log\log X + 0.281 - \sum_{p\leqslant 79}\frac{1}{p} < \log\log X - 1.488. \tag{10.35}$$

On collecting together (10.30), (10.31) and (10.33)–(10.35), we arrive at the upper bound
$$\prod_{7\leqslant p\leqslant X}(1+H(p)) < 4.35\cdot 3.88\cdot \exp\left(-1.488\cdot 3 + \frac{6.171}{316} + \frac{85.16}{340}\right)(\log X)^3$$
$$< 0.2547(\log X)^3. \tag{10.36}$$

On recalling (10.25) and (10.28), we deduce from (10.36) that
$$\prod_{p\leqslant X}(1+H(p)) < \frac{28129}{640}(1+2.57)(1+2.123)\cdot 0.2547(\log X)^3$$
$$< (5\log X)^3,$$

and hence the first conclusion of the lemma follows from (10.22).

We now consider the final assertion of the lemma. Observe first that by Lemma 10.2 (ii) one has $r(2) > 2^3$, and similarly, by Lemmata 10.2 (i) and 10.1 (ii), one has $r(p) \geqslant p^3$ for every odd prime p. Since $r(d) \geqslant 0$ for all natural numbers d, we find that the hypotheses necessary for the application of Lemma 10.4 are satisfied with $\lambda(d) = r(d)$. On recalling (10.21), we deduce that the final conclusion of the lemma follows from the proposition that for $X \geqslant 10^{25}$, one has

$$\prod_{p \leqslant X} \left(1 - \frac{1}{p} + H(p)\right) < (16 \log X)^2. \tag{10.37}$$

Fortunately, much of the work required to establish (10.37) has already been completed earlier in this proof. Observe first that, as in the argument leading from (10.29) to (10.30) and (10.31), a direct computation demonstrates that

$$\prod_{\substack{7 \leqslant p \leqslant 79 \\ p \equiv 3 \pmod 4}} \left(1 - \frac{1}{p} + H(p)\right) < 2.985, \tag{10.38}$$

and

$$\prod_{\substack{13 \leqslant p \leqslant 73 \\ p \equiv 1 \pmod 4}} \left(1 - \frac{1}{p} + H(p)\right) < 3.120. \tag{10.39}$$

Also, as in the argument leading to (10.32), we obtain for $p \geqslant 83$ the inequality

$$1 - \frac{1}{p} + H(p) < 1 + \frac{2}{p} + (3.038 b_p^3 + 3.133) p^{-2}$$

$$< \exp\left(\frac{2}{p} + (3.038 b_p^3 + 3.133) p^{-2}\right).$$

Then by (10.34), (10.35), (10.38) and (10.39), we deduce that for $X \geqslant 10^{25}$,

$$\prod_{7 \leqslant p \leqslant X} \left(1 - \frac{1}{p} + H(p)\right)$$

$$< 2.985 \cdot 3.120 \exp\left(2(\log \log X - 1.488) + \frac{6.171}{316} + \frac{85.16}{340}\right)$$

$$< 0.6222 (\log X)^2. \tag{10.40}$$

Hence, by (10.25) and (10.28), we may finally conclude that

$$\prod_{p \leqslant X} \left(1 - \frac{1}{p} + H(p)\right)$$

$$< \left(\frac{28129}{640} - \frac{1}{2}\right)\left(1 - \frac{1}{3} + 2.57\right)\left(1 - \frac{1}{5} + 2.123\right) \cdot 0.6222 (\log X)^2$$

$$< 256 (\log X)^2,$$

and this confirms the estimate (10.37), thereby completing the proof of the final conclusion of the lemma. □

Our final lemma in this section provides an analogue of Lemma 10.5 in which $s(d; \zeta)$ is substituted for $r(d; \varepsilon)$. This lemma provides the crucial ingredient in our proof of Lemma 2.5.

CHAPTER 10. AN INVESTIGATION OF CERTAIN CONGRUENCES

LEMMA 10.6. — *Suppose that* $X \geqslant 10^{25}$ *and* $\zeta \in \{0,1\}$. *Then*
$$\sum_{d \leqslant X} g(d)s(d;\zeta)d^{-4} < 86.7(\log X)^3,$$
and, whenever $0 \leqslant k \leqslant 3$, *one has also*
$$\sum_{d \leqslant X} g(d)s(d;\zeta)d^{-k} < 183 X^{4-k}(\log X)^2.$$

Proof. — Let $\zeta \in \{0,1\}$, and write $s(d) = s(d;\zeta)$. Putting
$$H'(p) = \sum_{l=1}^{\infty} g(p^l)s(p^l)p^{-4l}, \tag{10.41}$$
it follows from the multiplicative properties of $g(d)$ and $s(d)$ that
$$\sum_{d \leqslant X} g(d)s(d)d^{-4} \leqslant \prod_{p \leqslant X}(1 + H'(p)). \tag{10.42}$$

The similarity between the formulae (10.21) and (10.41), and likewise between the inequalities (10.22) and (10.42), is suggestive of a strategy for proving Lemma 10.6 similar to that applied in the proof of Lemma 10.5. Fortunately, we may be economical in our account by recycling the estimates contained in the latter proof.

First, by Lemma 10.3 (iii) and the definition (9.2) of $g(d)$, we have
$$1 + H'(2) = 1 + 3 + 3 + 9\sum_{l=3}^{\infty} 2^{3l+2} \cdot 2^{-4l} + 18 \sum_{l \in \{7,9,11\}} 2^{3l+2} \cdot 2^{-4l}$$
$$= 7 + 9 + \frac{18 \cdot 21}{512} = \frac{4285}{256}. \tag{10.43}$$

For odd primes p, we begin by noting that when $p \geqslant 7$, it follows from Lemmata 10.2 (i) and 10.3 (i), together with a comparison of (10.21) and (10.41), that one has $H'(p) \leqslant H(p)$. When $p = 3$ or 5, meanwhile, one finds in a similar manner from Lemmata 10.2 (i) and 10.3 (ii) that $H'(p) < \frac{3}{2}H(p)$. We therefore infer from (10.28), (10.36), (10.40) and (10.43) that whenever $X \geqslant 10^{25}$, one has
$$\prod_{p \leqslant X}(1 + H'(p))$$
$$\leqslant \frac{4285}{256}\left(1 + \frac{3}{2} \cdot 2.57\right)\left(1 + \frac{3}{2} \cdot 2.123\right) \prod_{7 \leqslant p \leqslant X}(1 + H(p))$$
$$< 86.7(\log X)^3, \tag{10.44}$$

and also
$$\prod_{p \leqslant X}\left(1 - \frac{1}{p} + H'(p)\right) \leqslant \left(\frac{4285}{256} - \frac{1}{2}\right)\left(1 - \frac{1}{3} + \frac{3}{2} \cdot 2.57\right)$$
$$\times \left(1 - \frac{1}{5} + \frac{3}{2} \cdot 2.123\right) \prod_{7 \leqslant p \leqslant X}\left(1 - \frac{1}{p} + H(p)\right)$$
$$< 183(\log X)^2. \tag{10.45}$$

The first conclusion of the lemma is now immediate from (10.42) and (10.44). As for the second conclusion of the lemma, one finds that Lemma 10.3 confirms that $s(p) \geqslant p^3$ for every prime p, and also $s(d) \geqslant 0$ for all natural numbers d. Hence we may apply Lemma 10.4 to obtain the desired conclusion without more ado from (10.45). This completes the proof of the lemma. □

CHAPTER 11

MEAN VALUE ESTIMATES

The smell of victory now lies heavy in the air, so we pause no longer before wielding Lemma 10.5 to establish Lemma 2.4.

Proof of Lemma 2.4. — We convert the mean value central to Lemma 2.4 into a divisor sum to which the methods of §10 apply. When $\varepsilon \in \{0, 1\}$, define the polynomial $\psi(\boldsymbol{x}; \varepsilon)$ by

$$\psi(\boldsymbol{x}; \varepsilon) = \frac{1}{2}\big((2x_1 + \varepsilon)^4 + (2x_2 + \varepsilon)^4 - (2x_3 + \varepsilon)^4 - (2x_4 + \varepsilon)^4\big),$$

and note that $\psi(\boldsymbol{x}; \varepsilon)$ is a polynomial in $x_1, \ldots, x_4$ with integral coefficients. By orthogonality, the mean value

$$\int_0^1 |F_\eta(\alpha)^2 S_\varepsilon(\alpha)^4| d\alpha$$

is then equal to the number of solutions of the equation

$$m_1^2 - m_2^2 = \psi(\boldsymbol{x}; \varepsilon), \tag{11.1}$$

with

$$m_1, m_2 \in \mathcal{M}_\eta(P^2) \quad \text{and} \quad 2P < 2x_j + \varepsilon \leqslant 4P \quad (1 \leqslant j \leqslant 4). \tag{11.2}$$

We denote by V_0 the number of solutions $\boldsymbol{m}, \boldsymbol{x}$ of (11.1) subject to (11.2) that satisfy the additional condition $\psi(\boldsymbol{x}; \varepsilon) = 0$, and we denote by V_1 the corresponding number of solutions with $\psi(\boldsymbol{x}; \varepsilon) \neq 0$. Thus we have

$$\int_0^1 |F_\eta(\alpha)^2 S_\varepsilon(\alpha)^4| d\alpha = V_0 + V_1. \tag{11.3}$$

In order to estimate V_0, we observe that whenever $\varepsilon \in \{0, 1\}$ and $P \geqslant 10^{25}$, the upper bound (3.3.14) of Deshouillers and Dress [9] supplies the estimate

$$\int_0^1 |S_\varepsilon(\alpha)|^4 d\alpha \leqslant 60 P^2 (\log P)^4. \tag{11.4}$$

Some comments are in order at this point concerning the validity of the estimate (11.4), since Deshouillers and Dress claim such an upper bound only for $P \geqslant 10^{80}$. However, an inspection of the proof of Theorem 3 of [9] reveals that the latter hypothesis is employed in the proof of (3.3.14) of [9] only in the application of Proposition 3.1 of that paper. Moreover, the latter proposition requires the hypothesis $P \geqslant 10^{80}$ only in the application of the relation (3.18) of Rosser and Schoenfeld [19] to estimate the right hand side of (3.1.24) of [9]. But on substituting our Lemma 3.2 for this result of Rosser and Schoenfeld, one obtains the same conclusion as that required in [9]. Thus, in the notation employed in [9], one finds that when $X \geqslant 10^{25}$,

$$A(X) \leqslant 40X^2 \exp\left(4 \sum_{5 \leqslant p \leqslant X} \frac{1}{p} + 1\right)$$
$$\leqslant 40X^2 \exp(4 \log \log X - 1.209) < 12X^2(\log X)^4,$$

and this last bound matches that found on the bottom of p.135 of [9]. Returning to the estimation of V_0, we find that on considering the underlying diophantine equation, one has

$$V_0 = M_\eta \int_0^1 |S_\varepsilon(\alpha)|^4 d\alpha \leqslant 60 M_\eta P^2 (\log P)^4. \tag{11.5}$$

We turn now to the solutions $\boldsymbol{m}, \boldsymbol{x}$ of (11.1) counted by V_1. Here we note that the definition of $\mathcal{M}_\eta(P^2)$, together with the constraint on $\psi(\boldsymbol{x};\varepsilon)$ imposed by (11.1), implies that $|\psi(\boldsymbol{x};\varepsilon)| \leqslant P^4$. Furthermore, it is evident that $m_1 - m_2$ has the same sign as ψ, and further that $|m_1 - m_2| \leqslant m_1 + m_2$. We therefore deduce that for a fixed choice of $\boldsymbol{x}$ with $\psi(\boldsymbol{x};\varepsilon) \neq 0$, the number of pairs (m_1, m_2) satisfying (11.1) with $m_i \in \mathcal{M}_\eta(P^2)$ $(i = 1, 2)$ does not exceed $\frac{1}{2}\tau(|\psi(\boldsymbol{x};\varepsilon)|)$. Consequently, one has

$$V_1 \leqslant \frac{1}{2} \sum_{\substack{\boldsymbol{x} \\ 1 \leqslant |\psi(\boldsymbol{x};\varepsilon)| \leqslant P^4}} \tau(|\psi(\boldsymbol{x};\varepsilon)|),$$

where the sum is over integral 4-tuples $\boldsymbol{x}$ with $2P < 2x_j + \varepsilon \leqslant 4P$ $(1 \leqslant j \leqslant 4)$. On applying Lemma 9.10, we find that

$$V_1 \leqslant 4 \sum_{\substack{\boldsymbol{x} \\ 1 \leqslant |\psi(\boldsymbol{x};\varepsilon)| \leqslant P^4}} \sum_{\substack{d \mid \psi(\boldsymbol{x};\varepsilon) \\ d \leqslant |\psi(\boldsymbol{x};\varepsilon)|^{1/4}}} g(d)$$

$$\leqslant 4 \sum_{d \leqslant P} g(d) \sum_{\substack{\boldsymbol{x} \\ \psi(\boldsymbol{x};\varepsilon) \equiv 0 \pmod{d}}} 1.$$

But plainly,

$$\sum_{\substack{\boldsymbol{x} \\ \psi(\boldsymbol{x};\varepsilon) \equiv 0 \pmod{d}}} 1 = \sum_{\substack{1 \leqslant a_1, \ldots, a_4 \leqslant d \\ \psi(\boldsymbol{a};\varepsilon) \equiv 0 \pmod{d}}} \sum_{\substack{\boldsymbol{x} \\ \boldsymbol{x} \equiv \boldsymbol{a} \pmod{d}}} 1$$

$$\leqslant r(d;\varepsilon)(P/d + 1)^4,$$

where $r(d;\varepsilon)$ is defined as in the preamble to Lemma 10.2. We therefore deduce that
$$V_1 \leqslant 4 \sum_{k=0}^{4} \binom{4}{k} P^k \sum_{d \leqslant P} \frac{g(d)r(d;\varepsilon)}{d^k},$$
whence by Lemma 10.5, we conclude that whenever $P \geqslant 10^{25}$,
$$V_1 < 4P^4 \left((5 \log P)^3 + \sum_{k=0}^{3} \binom{4}{k} (16 \log P)^2 \right)$$
$$= 500 P^4 (\log P)^3 + 15360 P^4 (\log P)^2. \tag{11.6}$$

The conclusion of Lemma 2.4 follows on substituting (11.5) and (11.6) into (11.3). □

The argument required to establish Lemma 2.5 is modelled after that above, though one encounters some mild complications.

Proof of Lemma 2.5. — On recalling (2.1) and (2.7), one finds by orthogonality that the integral
$$\int_0^1 |D_\zeta(\alpha)^2 S_1(\alpha)^2| d\alpha$$
is bounded above by the number of solutions of the equation
$$4m_1^2 + 24m_1(2x_1 + \zeta)^2 + 6(2x_1 + \zeta)^4 + (2y_1 + 1)^4$$
$$= 4m_2^2 + 24m_2(2x_2 + \zeta)^2 + 6(2x_2 + \zeta)^4 + (2y_2 + 1)^4, \tag{11.7}$$
with
$$m_1, m_2 \in \mathcal{M}_0(3P^2/7), \quad 1 \leqslant x_1, x_2 < P/6, \quad P \leqslant y_1, y_2 < 2P. \tag{11.8}$$
On putting
$$\phi(\boldsymbol{x}, \boldsymbol{y}; \zeta) = \frac{1}{4}\big(30((2x_1 + \zeta)^4 - (2x_2 + \zeta)^4) - ((2y_1 + 1)^4 - (2y_2 + 1)^4)\big),$$
we may rewrite the equation (11.7) as
$$(m_1 + 3(2x_1 + \zeta)^2)^2 - (m_2 + 3(2x_2 + \zeta)^2)^2 = \phi(\boldsymbol{x}, \boldsymbol{y}; \zeta). \tag{11.9}$$
Notice here that $\phi(\boldsymbol{x}, \boldsymbol{y}; \zeta)$ is a polynomial in $\boldsymbol{x}$ and $\boldsymbol{y}$ with integral coefficients. We denote by W_0 the number of solutions $\boldsymbol{m}, \boldsymbol{x}, \boldsymbol{y}$ of (11.9) subject to (11.8) that satisfy the additional condition that $\phi(\boldsymbol{x}, \boldsymbol{y}; \zeta) = 0$, and we denote by W_1 the corresponding number of solutions with $\phi(\boldsymbol{x}, \boldsymbol{y}; \zeta) \neq 0$. Thus we find that
$$\int_0^1 |D_\zeta(\alpha)^2 S_1(\alpha)^2| d\alpha = W_0 + W_1. \tag{11.10}$$

We begin by examining W_0, noting that for each fixed choice of $\boldsymbol{x}$ and $\boldsymbol{y}$ satisfying $\phi(\boldsymbol{x}, \boldsymbol{y}; \zeta) = 0$ and (11.8), it follows from (11.8) and (11.9) that the variables m_1 and m_2 satisfy
$$m_1 = m_2 + 3(2x_2 + \zeta)^2 - 3(2x_1 + \zeta)^2.$$

On recalling (2.13), we find that there are $\widetilde{M}_0$ such choices available for m_1 and m_2, whence
$$W_0 \leqslant \widetilde{M}_0 \int_0^1 |\widetilde{S}_\zeta(30\alpha)^2 S_1(\alpha)^2| d\alpha, \tag{11.11}$$
where we write
$$\widetilde{S}_\zeta(\alpha) = \sum_{1 \leqslant x < P/6} e((2x+\zeta)^4 \alpha).$$
We claim that whenever $\zeta \in \{0, 1\}$ and $P \geqslant 10^{26}$, then one has
$$\int_0^1 |\widetilde{S}_\zeta(30\alpha)|^4 d\alpha \leqslant 60(P/6)^2 \big(\log(P/6)\big)^4. \tag{11.12}$$
In order to confirm this upper bound, we first observe that a change of variables yields
$$\int_0^1 |\widetilde{S}_\zeta(30\alpha)|^4 d\alpha = \int_0^1 |\widetilde{S}_\zeta(\alpha)|^4 d\alpha,$$
and that by orthogonality, the latter integral can be seen to count the number of solutions of the same diophantine equation as that underlying the left hand side of (11.4), save that the variables now lie in the interval $[1, P/6)$ as opposed to $(P-\varepsilon/2, 2P-\varepsilon/2]$. On examining equation (3.1.2) of Deshouillers and Dress [9], however, we find that the upper bound for $A(X)$ concluding p.135 of [9] holds when the implicit interval $(X, 2X]$ of [9] is replaced by $[1, X)$, and thus our earlier discussion pertaining to (11.4) remains valid in the current situation. The desired conclusion (11.12) therefore holds whenever $P/6 \geqslant 10^{25}$. Finally, on combining (11.4), (11.11) and (11.12) via Schwarz's inequality, we conclude that
$$W_0 \leqslant \widetilde{M}_0 \left(\int_0^1 |\widetilde{S}_\zeta(30\alpha)|^4 d\alpha \right)^{1/2} \left(\int_0^1 |S_1(\alpha)|^4 d\alpha \right)^{1/2}$$
$$\leqslant 10 \widetilde{M}_0 P^2 (\log P)^4. \tag{11.13}$$

Next we turn to W_1. Let W_2 be the number of solutions of the equation
$$n_1^2 - n_2^2 = \phi(\boldsymbol{x}, \boldsymbol{y}; \zeta), \tag{11.14}$$
with $\boldsymbol{x}$ and $\boldsymbol{y}$ satisfying the conditions recorded in (11.8), and with $1 \leqslant n_1, n_2 \leqslant P^2$ and $\phi(\boldsymbol{x}, \boldsymbol{y}; \zeta) \neq 0$. When $\boldsymbol{m}$ and $\boldsymbol{x}$ satisfy (11.8), one has
$$1 \leqslant m_j + 3(2x_j + \zeta)^2 < P^2 \qquad (j=1, 2),$$
and so it is apparent that
$$W_1 \leqslant W_2. \tag{11.15}$$

Let $\boldsymbol{n}, \boldsymbol{x}, \boldsymbol{y}$ be a solution of (11.14) counted by W_2. Then it follows that $|\phi(\boldsymbol{x}, \boldsymbol{y}; \zeta)| \leqslant P^4$. Furthermore, on following the argument of the proof of Lemma 2.4 above, we find that the number of pairs (n_1, n_2) satisfying (11.14) with $1 \leqslant n_i \leqslant P^2$ ($i = 1, 2$) does not exceed $\frac{1}{2}\tau(|\phi(\boldsymbol{x}, \boldsymbol{y}; \zeta)|)$. We thus deduce that

$$W_2 \leqslant \frac{1}{2} \sum_{\substack{\boldsymbol{x}, \boldsymbol{y} \\ 1 \leqslant |\phi(\boldsymbol{x}, \boldsymbol{y}; \zeta)| \leqslant P^4}} \tau(|\phi(\boldsymbol{x}, \boldsymbol{y}; \zeta)|),$$

where the sum is over integral 4-tuples $(\boldsymbol{x}, \boldsymbol{y})$ with $1 \leqslant x_j < P/6$ and $P \leqslant y_j < 2P$ ($j = 1, 2$). On applying Lemma 9.10, we obtain

$$W_2 \leqslant 4 \sum_{\substack{\boldsymbol{x}, \boldsymbol{y} \\ 1 \leqslant |\phi(\boldsymbol{x}, \boldsymbol{y}; \zeta)| \leqslant P^4}} \sum_{\substack{d | \phi(\boldsymbol{x}, \boldsymbol{y}; \zeta) \\ d \leqslant |\phi(\boldsymbol{x}, \boldsymbol{y}; \zeta)|^{1/4}}} g(d)$$

$$\leqslant 4 \sum_{d \leqslant P} g(d) \sum_{\substack{\boldsymbol{x}, \boldsymbol{y} \\ \phi(\boldsymbol{x}, \boldsymbol{y}; \zeta) \equiv 0 \pmod d}} 1.$$

But

$$\sum_{\substack{\boldsymbol{x}, \boldsymbol{y} \\ \phi(\boldsymbol{x}, \boldsymbol{y}; \zeta) \equiv 0 \pmod d}} 1 = \sum_{\substack{1 \leqslant a_1, a_2 \leqslant d \\ 1 \leqslant b_1, b_2 \leqslant d \\ \phi(\boldsymbol{a}, \boldsymbol{b}; \zeta) \equiv 0 \pmod d}} \sum_{\substack{\boldsymbol{x}, \boldsymbol{y} \\ \boldsymbol{x} \equiv \boldsymbol{a} \pmod d \\ \boldsymbol{y} \equiv \boldsymbol{b} \pmod d}} 1$$

$$\leqslant s(d; \zeta) \left(\frac{P}{6d} + 1\right)^2 \left(\frac{P}{d} + 1\right)^2,$$

where $s(d; \zeta)$ is defined as in the preamble to Lemma 10.3. We therefore find that

$$W_2 \leqslant 4\left(\frac{1}{36}\Sigma_4 + \frac{7}{18}\Sigma_3 + \frac{61}{36}\Sigma_2 + \frac{7}{3}\Sigma_1 + \Sigma_0\right),$$

where

$$\Sigma_k = \sum_{d \leqslant P} g(d) s(d; \zeta) d^{-k} \qquad (0 \leqslant k \leqslant 4).$$

Consequently, on making use of Lemma 10.6, we conclude that whenever $P \geqslant 10^{50}$, one has

$$W_2 \leqslant 4\left(\frac{86.7}{36} + \left(\frac{7}{18} + \frac{61}{36} + \frac{7}{3} + 1\right)\frac{183}{\log P}\right) P^4 (\log P)^3$$

$$< 44.1 P^4 (\log P)^3. \tag{11.16}$$

The conclusion of the lemma follows on substituting (11.13) and (11.16) into (11.15) and (11.10). $\square$

CHAPTER 12

APPENDIX: SUMS OF NINETEEN BIQUADRATES

In this section we describe our proof that $g(4) = 19$, which is to say that every natural number is a sum of at most nineteen biquadrates. The programme of Balasubramanian, Deshouillers and Dress for proving that $g(4) = 19$, based on Hua's inequality, sought to establish that every integer exceeding 10^{367} is a sum of nineteen biquadrates. By virtue of the tools prepared in previous sections, we are now able to reduce the above limit 10^{367} substantially, thereby easing the computational burden of showing that $g(4) = 19$.

THEOREM 3. — *Every integer exceeding 10^{146} can be written as the sum of nineteen biquadrates.*

Proof. — According to Lemma 4.2, we may take a real number ν with

$$53 < \nu < 107.5, \tag{12.1}$$

such that whenever $\nu - 1/4 \leqslant \xi \leqslant \nu$, one has

$$K_{13}(\xi) > 0.0065865. \tag{12.2}$$

Fixing such a real number ν, we consider a large natural number N, and define the positive numbers P_0 and P by means of the relations (2.10). Further, when $N \equiv r \pmod{16}$ with $1 \leqslant r \leqslant 16$, we define the integers η and t by

$$\begin{cases} \eta = 0 \quad \text{and} \quad t = r, & \text{for } 1 \leqslant r \leqslant 4, \\ \eta = 1 \quad \text{and} \quad t = r - 4, & \text{for } 5 \leqslant r \leqslant 16. \end{cases}$$

We note that in all circumstances, our choices for η and t ensure that

$$1 \leqslant t \leqslant 12 \quad \text{and} \quad N - 4\eta \equiv t \pmod{16}. \tag{12.3}$$

With the above conventions in hand, we denote by $R'(N)$ the number of representations of N in the form

$$N = 2m_1^2 + 2m_2^2 + \sum_{j=1}^{13-t}(2x_j)^4 + \sum_{l=1}^{t}(2y_l+1)^4,$$

subject to

$$m_1, m_2 \in \mathcal{M}_\eta(P^2), \quad P < x_j \leqslant 2P \ (1 \leqslant j \leqslant 13-t), \quad P \leqslant y_l < 2P \ (1 \leqslant l \leqslant t).$$

Thus, in view of the identity (1.1), together with the definition of the sets $\mathcal{M}_\eta(X)$ from §2, it follows that whenever $R'(N) > 0$, then N can be written as a sum of nineteen biquadrates.

We next recall the definitions of $S_\varepsilon(\alpha)$, $F_\eta(\alpha)$, $\mathfrak{U}$, $\mathfrak{M}$ and $\mathfrak{m}$ (see, especially, equations (2.1), (2.2) and (2.6)). Also, when $\mathfrak{L} \subseteq \mathfrak{U}$, we define $R'(N;\mathfrak{L})$ by

$$R'(N;\mathfrak{L}) = \int_\mathfrak{L} F_\eta(\alpha)^2 S_0(\alpha)^{13-t} S_1(\alpha)^t e(-N\alpha) d\alpha,$$

so that

$$R'(N) = R'(N;\mathfrak{U}) = R'(N;\mathfrak{M}) + R'(N;\mathfrak{m}). \tag{12.4}$$

One may estimate $R'(N;\mathfrak{M})$ in a straightforward manner by following the treatment that we applied in the proof of Lemma 2.1. We first estimate the integral

$$\Phi'(n;t) = \int_\mathfrak{M} S_0(\alpha)^{13-t} S_1(\alpha)^t e(-n\alpha) d\alpha$$

for natural numbers n with $N - 4P^4 \leqslant n \leqslant N$ and $n \equiv t \pmod{16}$. Recalling the notation adopted in the proof of Lemma 7.2, we write

$$\Phi'_1(n;t) = \sum_{q \leqslant P^{1/2}} \sum_{\substack{a=1 \\ (a,q)=1}}^{q} \int_{|\beta| \leqslant 975(qP^3)^{-1}} T_0^{13-t} T_1^t e(-(a/q+\beta)n) d\beta,$$

$$\Phi'_{1,1} = \sum_{q \leqslant P^{1/2}} \sum_{\substack{a=1 \\ (a,q)=1}}^{q} \int_0^\infty |T_0|^{12} U d\beta, \qquad \Phi'_{1,2} = \frac{975}{P^3} \sum_{q \leqslant P^{1/2}} U^{13}.$$

Then, on imitating the derivation of the estimate (7.3), we find that

$$|\Phi'(n;t) - \Phi'_1(n;t)| \leqslant 2^{14}(\Phi'_{1,1} + \Phi'_{1,2}). \tag{12.5}$$

In order to estimate $\Phi'_{1,1}$, we define

$$V'(q) = \sum_{\substack{a=1 \\ (a,q)=1}}^{q} |q^{-1} G_0(q,a)|^{12} \quad \text{and} \quad W'(p) = \sum_{l=0}^{\infty} p^{l/4} V'(p^l),$$

and observe that $V'(q)$ is multiplicative. From (7.6) it follows that

$$\int_0^\infty |I(\beta)|^{12} d\beta \leqslant \frac{3}{11\pi}(2P_0)^8,$$

and thus one obtains the upper bound

$$\Phi'_{1,1} \leqslant \frac{3}{11\pi}(2P_0)^8 \times 3 \times 10^6 \times 2^{-12}P^{1/2} \sum_{q \leqslant P^{1/2}} q^{1/4}V'(q)$$
$$< 16278 P_0^{8.5} \prod_p W'(p). \tag{12.6}$$

An application of Lemma 5.2 demonstrates, via a direct computation, that

$$W'(2) \leqslant 1 + \sum_{l=1}^{4} 2^{5l/4-1} + \sum_{u=1}^{\infty} \sum_{v=1}^{4} 2^{5(4u+v)/4-1} \kappa(2^{4u+v})^{12}$$
$$= 1 + \sum_{l=1}^{4} 2^{5l/4-1} + \sum_{u=1}^{\infty} 2^{-7u-1} \sum_{v=1}^{4} 2^{5v/4} c(2^v)^{12}$$
$$< 522. \tag{12.7}$$

For odd primes p, one deduces from (5.3), (5.8) and Lemmata 5.2 and 5.3 that for $u \geqslant 0$, one has

$$V'(p^{4u+1}) = p^{-8u-12} \sum_{a=1}^{p-1} |S(p,a)|^{12}$$
$$\leqslant p^{-8u-12}(p\kappa(p))^{10} \sum_{a=1}^{p-1} |S(p,a)|^{2}$$
$$= b_p p^{-8u-1}(p-1)\kappa(p)^{10}.$$

Also, for $u \geqslant 0$ and $2 \leqslant v \leqslant 4$, it follows from Lemma 5.2 together with (5.6) that

$$V'(p^{4u+v}) \leqslant p^{-8u+v-13}(p-1).$$

Consequently, we obtain

$$W'(p) \leqslant 1 + (p-1) \sum_{u=0}^{\infty} \left(b_p p^{-7u-3/4} \kappa(p)^{10} + \sum_{v=2}^{4} p^{-7u+5v/4-13} \right)$$
$$= 1 + \frac{p-1}{1-p^{-7}} \left(b_p \kappa(p)^{10} p^{-3/4} + p^{-13}(p^{5/2} + p^{15/4} + p^5) \right)$$
$$< 1 + b_p \kappa(p)^{10} p^{1/4} + 3p^{-7}.$$

A direct computation now reveals that

$$W'(3) < 1.00679 \quad \text{and} \quad \prod_{\substack{p \leqslant 73 \\ p \equiv 1 \pmod 4}} W'(p) < 2.513.$$

Similarly, on recalling Lemma 5.4, one finds that

$$\prod_{\substack{p \geqslant 7 \\ p \equiv 3 \pmod 4}} W'(p) < \exp\left(\sum_{\substack{p \geqslant 7 \\ p \equiv 3 \pmod 4}} (p^{-19/4} + 3p^{-7}) \right) < 1.00126$$

and
$$\prod_{\substack{p \geqslant 89 \\ p \equiv 1 \,(\text{mod } 4)}} W'(p) < \exp\left(\sum_{\substack{p \geqslant 89 \\ p \equiv 1 \,(\text{mod } 4)}} (3^{11} p^{-19/4} + 3p^{-7})\right) < 1.00069.$$

Combining these estimates with the upper bounds (12.6) and (12.7), we conclude that
$$\Phi'_{1,1} < 2.155 \times 10^7 P_0^{8.5}. \tag{12.8}$$

Turning next to the estimation of $\Phi'_{1,2}$, we find by a simple calculation that
$$\Phi'_{1,2} = 975 \times 3^{13} \times 10^{78} P^{7/2} \sum_{q \leqslant P^{1/2}} q^{13/4}$$
$$\leqslant 975 \times 3^{13} \times 10^{78} P^{45/8}. \tag{12.9}$$

Thus it follows from (12.5), (12.8) and (12.9) that whenever $P \geqslant 10^{35}$, one has
$$|\Phi'(n;t) - \Phi'_1(n;t)| < 1.2 \times 10^{-6} P^9. \tag{12.10}$$

We next define
$$\Phi'_2(n;t) = \sum_{q \leqslant P^{1/2}} \sum_{\substack{a=1 \\ (a,q)=1}}^{q} \int_{-\infty}^{\infty} T_0^{13-t} T_1^t e(-(a/q + \beta)n) d\beta.$$

By (7.6) and the trivial bound $|G_\varepsilon(q,a)| \leqslant q$, one easily obtains the estimate
$$|\Phi'_1(n;t) - \Phi'_2(n;t)| \leqslant \sum_{q \leqslant P^{1/2}} \sum_{a=1}^{q} 2 \int_{975(qP^3)^{-1}}^{\infty} 2^{-13} |I(\beta)|^{13} d\beta$$
$$< P^4. \tag{12.11}$$

Meanwhile, on writing
$$A'(q,n;t) = q^{-13} \sum_{\substack{a=1 \\ (a,q)=1}}^{q} G_0(q,a)^{13-t} G_1(q,a)^t e(-an/q)$$

and
$$\mathfrak{S}'(n,Q;t) = \sum_{q \leqslant Q} A'(q,n;t),$$

a suitable change of variables leads from (7.5) and (4.2) to the conclusion
$$\Phi'_2(n;t) = \mathfrak{S}'(n, P^{1/2}; t) K_{13}(n/(16P_0^4)) P_0^9 / 16. \tag{12.12}$$

By Lemma 5.2 together with (5.7), one has the bound
$$|A'(q,n;t)| \leqslant q^{-12} (q\kappa(q))^{13} < 9^{13} q^{-9/4},$$

and this assures the absolute convergence of the infinite series $\mathfrak{S}'(n;t)$ defined by
$$\mathfrak{S}'(n;t) = \sum_{q=1}^{\infty} A'(q,n;t).$$

Furthermore, when $P \geqslant 10^{35}$, we obtain the estimate
$$|\mathfrak{S}'(n;t) - \mathfrak{S}'(n, P^{1/2};t)| \leqslant 9^{13} \int_{P^{1/2}-1}^{\infty} z^{-9/4} dz < 10^{-8}. \tag{12.13}$$

A lower bound for $\mathfrak{S}'(n;t)$ can be established by the methods described in §6. We define
$$B'(p, n; t) = \sum_{l=0}^{\infty} A'(p^l, n; t),$$
and note that $\mathfrak{S}'(n;t)$ may be written as the infinite product
$$\mathfrak{S}'(n;t) = \prod_{p} B'(p, n; t).$$

One may estimate $B'(p, n; t)$ by considering the number of solutions of an associated congruence. Thus, arguing as in the derivation of (6.6) above, it is swiftly confirmed that when $n \equiv t \pmod{16}$, one has
$$B'(2, n; t) = \sum_{l=0}^{4} A'(2^l, n; t) = 16. \tag{12.14}$$

Also, the argument leading to (6.9) shows that for $p = 3$ and 5,
$$B'(p, n; t) \geqslant p^{-12} \min_{0 \leqslant r \leqslant p-1} \left\{ \sum_{\substack{1 \leqslant s \leqslant 13 \\ s \equiv r \pmod{p}}} \binom{13}{s}(p-1)^s \right\},$$
from which one derives the lower bounds
$$B'(3, n; t) \geqslant \frac{728}{729} \quad \text{and} \quad B'(5, n; t) \geqslant \frac{1211776}{1953125}. \tag{12.15}$$

Moreover, by following the argument of the proof of (6.17), one deduces from Lemmata 5.2 and 5.3 that for odd primes p, one has
$$B'(p, n; t) \geqslant 1 - b_p(1 - p^{-1})\kappa(p)^{11} - p^{-12}.$$

Using the last inequality, a modicum of computation confirms that
$$\prod_{\substack{13 \leqslant p \leqslant 73 \\ p \equiv 1 \pmod 4}} B'(p, n; t) > 0.9732, \tag{12.16}$$
and, with the aid of Lemma 5.4, one obtains the additional lower bounds
$$\prod_{\substack{p \geqslant 7 \\ p \equiv 3 \pmod 4}} B'(p, n; t) > \prod_{\substack{p \geqslant 7 \\ p \equiv 3 \pmod 4}} \exp(-p^{-11/2}) > 0.9996, \tag{12.17}$$
and
$$\prod_{\substack{p \geqslant 89 \\ p \equiv 1 \pmod 4}} B'(p, n; t) > \prod_{\substack{p \geqslant 89 \\ p \equiv 1 \pmod 4}} \exp(-3^{12} p^{-11/2}) > 0.9999. \tag{12.18}$$

Therefore, on combining (12.14)–(12.18), we deduce that when $n \equiv t \pmod{16}$, one has
$$\mathfrak{S}'(n;t) = \prod_p B'(p,n;t) > 9.6427. \tag{12.19}$$

Our forces are now poised for victory in this first substantial phase of our argument. Collecting together (12.2), (12.10)–(12.13), (12.19), and recalling (2.10), we conclude at this point that when
$$N - 4P^4 \leqslant n \leqslant N, \quad n \equiv t \pmod{16} \quad \text{and} \quad P \geqslant 10^{35},$$
one has
$$\Phi'(n;t) \geqslant (9.6427 - 10^{-8}) \times 0.0065865 P_0^9/16 - P^4 - 1.2 \times 10^{-6} P^9$$
$$> 0.003968 P^9. \tag{12.20}$$

The lower bound (12.20) provides a major arc estimate for an auxiliary problem involving only 13 biquadrates. We now apply this bound to obtain a lower bound for the major arc contribution $R'(N;\mathfrak{M})$ relevant to the problem central to this section. Observe that
$$R'(N;\mathfrak{M}) = \sum_{m_1, m_2 \in \mathcal{M}_\eta(P^2)} \Phi'(N - 2m_1^2 - 2m_2^2; t).$$
When $m_1, m_2 \in \mathcal{M}_\eta(P^2)$, it follows from (2.3) and (12.3) that
$$N - 2m_1^2 - 2m_2^2 \equiv t \pmod{16},$$
and it is also apparent that
$$N - 4P^2 \leqslant N - 2m_1^2 - 2m_2^2 \leqslant N.$$
Hence, on recalling the notation introduced in (2.13), we deduce from (12.20) that whenever $P \geqslant 10^{35}$, one has
$$R'(N;\mathfrak{M}) > 0.003968 M_\eta^2 P^9. \tag{12.21}$$

It remains for us to estimate the minor arc contribution $R'(N;\mathfrak{m})$. On writing
$$S(\alpha) = \max\{|S_0(\alpha)|, |S_1(\alpha)|\},$$
we find that
$$R'(N;\mathfrak{m}) \leqslant \left(\sup_{\alpha \in \mathfrak{m}} S(\alpha)\right)^9 \int_0^1 |F_\eta(\alpha)^2 S_\varepsilon(\alpha)^4| d\alpha,$$
where ε is 0 or 1 according to whether $1 \leqslant t \leqslant 3$ or $4 \leqslant t \leqslant 12$. Following the argument leading to (2.14), but assuming now that $P \geqslant 10^{35}$, we deduce from Lemmata 2.2 and 2.4 that
$$\int_0^1 |F_\eta(\alpha)^2 S_\varepsilon(\alpha)^4| d\alpha < 61290 M_\eta^2 (\log P)^{9/2}.$$
Thus, on applying also Lemma 2.3, we conclude that for $P \geqslant 10^{35}$ one has
$$|R'(N;\mathfrak{m})| < 77^9 \times 61290 M_\eta^2 P^{7.956} (\log P)^{6.75}. \tag{12.22}$$

We return at last to (12.4), now combining (12.21) and (12.22) to deduce that for $P \geqslant 10^{35}$ one has
$$R'(N) \geqslant R'(N; \mathfrak{M}) - |R'(N; \mathfrak{m})| > 0.003968 M_\eta^2 P^9 (1 - E'),$$
where
$$E' = 1.47 \times 10^{24} P^{-1.044} (\log P)^{6.75}.$$
A modest calculation reveals that $E' < 1$ whenever $P \geqslant 3.28 \times 10^{35}$, and also, by (2.10) and (12.1), this condition on P is satisfied whenever $N \geqslant N_1$, where
$$N_1 = 16\nu(3.28 \times 10^{35} + 1)^4 < 2 \times 10^{145}.$$
We therefore conclude that $R'(N) > 0$ whenever $N \geqslant 2 \times 10^{145}$, and this suffices to establish the theorem. □

Equipped with the conclusion of Theorem 3, all that remains to confirm that $g(4) = 19$ is to check that every natural number not exceeding 10^{146} is a B_{19} (the reader may wish to recall our convention concerning the notation B_s described following the statement of Theorem 1). Although such a check is executed in the work of Deshouillers and Dress [10], we nonetheless present an account here in order to more clearly describe the extent to which heavy computations are required to establish that $g(4) = 19$. It transpires that the conclusions recorded in the next lemma are sufficient for our purposes, and one may check all of the conclusions of this lemma within a couple of hours on even a modest personal computer.

To facilitate our subsequent discussion, we define the set $\mathcal{R}(q; a)$ for integers q and a by
$$\mathcal{R}(q; a) = \{m \in \mathbb{Z} : m \geqslant 0, \ m \equiv a \pmod{q}\},$$
and, for $\varepsilon = 0$ or 1, we define the set $\mathcal{Q}_\varepsilon$ by
$$\mathcal{Q}_\varepsilon = \{m \in \mathbb{Z} : m \geqslant 0, \ m \equiv \varepsilon \pmod{2}, \ 5 \nmid m\}.$$
Thus we have
$$\mathcal{Q}_\varepsilon = \mathcal{R}(2; \varepsilon) \smallsetminus \mathcal{R}(5; 0),$$
and it follows, in particular, that whenever $m \in \mathcal{Q}_\varepsilon$ for $\varepsilon = 0, 1$, one has
$$m^4 \in \mathcal{R}(16; \varepsilon) \quad \text{and} \quad m^4 \in \mathcal{R}(5; 1). \tag{12.23}$$

LEMMA 12.1. — *One has the following conclusions:*

(i) *Every natural number not exceeding* 13792 *is a* B_{19}, *and every integer in the interval* $[13793, 50000]$ *is a* B_{16};

(ii) *Let $\mathcal{A}$ denote the set of B_2 numbers in $\mathcal{R}(80; 17) \cap [0, 600^4]$ defined by*
$$\mathcal{A} = \{a^4 + b^4 : a \in \mathcal{Q}_0, b \in \mathcal{Q}_1\} \cap [0, 600^4],$$
and suppose that $\mathcal{A} = \{a_1, a_2, \cdots\}$, where $a_1 < a_2 < \cdots$. Then one has $a_1 = 17$, $a_{53401} = 129598530097$, and moreover $a_{j+1} - a_j \leqslant 40587360$ for $1 \leqslant j \leqslant 53400$;

(iii) *Every integer in $[1143331, 70^4 + 71^4] \cap \mathcal{R}(80; 51)$ is a B_6.*

Proof. — The authors confirmed this lemma by using the software package Mathematica on a standard computer with 32MB of RAM and CPU speed 150MHz. Making use of unsophisticated programs, parts (i) and (ii) of the lemma were verified within twenty minutes, and two minutes, respectively. Part (iii) of the lemma is the most difficult to verify, and for this we proceed as follows.

Let $\mathcal{B}$ denote the set of B_5 numbers in $[0, 70^4]$ that are of the form $b_1^4 + \cdots + b_5^4$, with $b_1, b_2, b_3 \in \mathcal{Q}_0$ and $b_4, b_5 \in \mathcal{Q}_1$. Note that $\mathcal{B} \subset \mathcal{R}(80; 50)$, and put

$$\mathcal{C} = [0, 70^4] \cap (\mathcal{R}(80; 50) \smallsetminus \mathcal{B}).$$

A simple computer program may be used to determine all of the elements of $\mathcal{C}$, and indeed the machine applied by the authors required fifty minutes to complete this task (we remark that $\mathrm{card}(\mathcal{C}) = 19687$).

Next, when $\mathcal{D}$ is a finite subset of $\mathcal{Q}_1$, we define

$$\mathcal{C}(\mathcal{D}) = \bigcap_{d \in \mathcal{D}} \{c + d^4 : c \in \mathcal{C}\}$$

and

$$I(\mathcal{D}) = \left(\bigcap_{d \in \mathcal{D}} [d^4, 70^4 + d^4] \right) \cap \mathcal{R}(80; 51).$$

Whenever $m \in I(\mathcal{D})$ and $d \in \mathcal{D} \subset \mathcal{Q}_1$, it follows from (12.23) that the integer $m - d^4$ belongs to the set $[0, 70^4] \cap \mathcal{R}(80; 50)$. Consequently, one finds that if $m \in I(\mathcal{D})$ and $m \notin \mathcal{C}(\mathcal{D})$, then for some $b \in \mathcal{B}$ and $d \in \mathcal{D}$ one has $m - d^4 = b$, whence $m = b + d^4$ is a B_6. We next define the sets

$$\mathcal{D}_1 = [1, 31] \cap \mathcal{Q}_1, \qquad \mathcal{D}_2 = [37, 49] \cap \mathcal{Q}_1, \qquad \mathcal{D}_3 = [59, 71] \cap \mathcal{Q}_1,$$
$$\mathcal{D}_4 = [63, 77] \cap \mathcal{Q}_1, \qquad \mathcal{D}_5 = [67, 79] \cap \mathcal{Q}_1, \qquad \mathcal{D}_6 = [71, 81] \cap \mathcal{Q}_1.$$

A straightforward computational check confirms that $\mathcal{C}(\mathcal{D}_1) = \{1143251\}$, and also that $\mathcal{C}(\mathcal{D}_j)$ is empty for each j with $2 \leqslant j \leqslant 6$. This task expended only a few minutes work on the computer employed by the authors. Accordingly, we recognise that if

$$m \in \bigcup_{j=1}^{6} I(\mathcal{D}_j) = [31^4, 70^4 + 71^4] \cap \mathcal{R}(80; 51)$$

and $m > 1143251$, then m is a B_6. This completes our account of part (iii) of the lemma. □

Our commitment of computational time at this point amounts to less than 90 minutes. We next employ Lemma 12.1 within the ascent arguments of Deshouillers and Dress [10], though we incorporate several minor modifications. We begin by extracting the following conclusion from Lemma 12.1 (ii) and (iii).

LEMMA 12.2. — *Every integer in* $[1143348, 1.2964 \times 10^{11}] \cap \mathcal{R}(80; 68)$ *is a* B_8.

Proof. — Recalling the integers a_j introduced in the statement of Lemma 12.1 (ii), it follows from Lemma 12.1 (iii) that for each index j, every integer in
$$[1143331 + a_j, \ 70^4 + 71^4 + a_j] \cap \mathcal{R}(80; 68)$$
is a B_8. But from Lemma 12.1 (ii), one finds that whenever $1 \leqslant j \leqslant 53400$, one has
$$70^4 + 71^4 + a_j \geqslant 70^4 + 71^4 - 40587360 + a_{j+1} > 1143331 + a_{j+1},$$
whence
$$\bigcup_{j=1}^{53401} [1143331 + a_j, \ 70^4 + 71^4 + a_j] = [1143348, \ 70^4 + 71^4 + a_{53401}].$$
The lemma follows immediately on noting that $70^4 + 71^4 + a_{53401} > 1.2964 \times 10^{11}$. □

Other ascent procedures are based on a simple fact that we record as the following lemma.

LEMMA 12.3. — *Let h be a natural number, and let x and y be real numbers with $y \geqslant 2h^4$. Suppose that $\mathcal{H}$ is a set of non-negative integers satisfying the property that amongst any h consecutive non-negative integers, at least one belongs to $\mathcal{H}$. Then whenever*
$$n \in [x + (h-1)^4, \ x + h^4 + (y/(4h))^{4/3}],$$
there exists an integer $m \in \mathcal{H}$ such that $n - m^4 \in [x, \ x+y]$.

Proof. — Put $y_h = (y/(4h))^{1/3}$, and write $\mathcal{H} \cap [0, y_h] = \{h_0, h_1, \cdots, h_k\}$ with $h_0 < h_1 < \cdots < h_k$. By assumption, we have
$$0 \leqslant h_0 \leqslant h - 1 \quad \text{and} \quad y_h - h \leqslant h_k \leqslant y_h, \qquad (12.24)$$
and, moreover, whenever $1 \leqslant j \leqslant k$, one has
$$(n - h_{j-1}^4) - (n - h_j^4) = h_j^4 - h_{j-1}^4 \leqslant h_j^4 - (h_j - h)^4 \leqslant 4hh_j^3 \leqslant 4hy_h^3 = y.$$
Thus, if $n - h_0^4 \geqslant x$ and $n - h_k^4 \leqslant x + y$, then there is an index j with $0 \leqslant j \leqslant k$ such that $n - h_j^4 \in [x, \ x+y]$. But by (12.24) we have $h_0^4 \leqslant (h-1)^4$, and provided that $y_h \geqslant 2h/3$, or equivalently $y \geqslant (32/27)h^4$, one has
$$y + h_k^4 \geqslant y + (y_h - h)^4 = y_h^4 + 6h^2 y_h^2 - 4h^3 y_h + h^4 \geqslant y_h^4 + h^4.$$
We therefore see that
$$[x + h_0^4, \ x + y + h_k^4] \supset [x + (h-1)^4, \ x + h^4 + y_h^4],$$
and the conclusion of the lemma now follows. □

LEMMA 12.4. — *Let x and y be real numbers satisfying $x \geqslant 0$ and $y \geqslant 20000$, let k and l be integers, and let s be an integer exceeding 1. Then the following conclusions hold.*

(i) *Suppose that every integer in $[x, \ x+y]$ is a B_s. Then every integer in $[x, \ x + (y/4)^{4/3}]$ is a B_{s+1}.*

(ii) *Suppose that every integer in* $[x, x+y] \cap \mathcal{R}(16;k)$ *is a* B_s. *Then every integer in*

$$[x+1, x+1+(y/8)^{4/3}] \cap \bigl(\mathcal{R}(16;k) \cup \mathcal{R}(16;k+1)\bigr)$$

is a B_{s+1}.

(iii) *Suppose that every integer in* $[x, x+y] \cap \mathcal{R}(16;k) \cap \mathcal{R}(5;l)$ *is a* B_s. *Then every integer in*

$$[x+81, x+81+(y/16)^{4/3}] \cap \bigl(\mathcal{R}(16;k) \cup \mathcal{R}(16;k+1)\bigr) \cap \mathcal{R}(5;l+1)$$

is a B_{s+1}.

(iv) *Under the same hypotheses as in case* (iii), *every integer in*

$$[x+6561, x+6561+(y/40)^{4/3}] \cap \bigl(\mathcal{R}(16;k) \cup \mathcal{R}(16;k+1)\bigr)$$
$$\cap \bigl(\mathcal{R}(5;l) \cup \mathcal{R}(5;l+1)\bigr)$$

is a B_{s+1}.

Proof. — By applying Lemma 12.3 with $\mathcal{H} = \mathcal{R}(1;0)$ and $h = 1$, we find that whenever $n \in [x, x+(y/4)^{4/3}]$, there exists a non-negative integer m for which $n - m^4 \in [x, x+y]$. The conclusion of part (i) follows immediately.

Next, by applying Lemma 12.3 with $\mathcal{H} = \mathcal{R}(2;\varepsilon)$ and $h = 2$, and recalling (5.11), we see that whenever

$$n \in [x+1, x+1+(y/8)^{4/3}] \cap \mathcal{R}(16;k+\varepsilon)$$

with $\varepsilon = 0$ or 1, there exists $m \in \mathcal{R}(2;\varepsilon)$ such that

$$n - m^4 \in [x, x+y] \cap \mathcal{R}(16;k).$$

This establishes part (ii) of the lemma.

By applying Lemma 12.3 with $\mathcal{H} = \mathcal{Q}_\varepsilon$ and $h = 4$, meanwhile, and recalling (12.23), we deduce that whenever

$$n \in [x+81, x+81+(y/16)^{4/3}] \cap \mathcal{R}(16;k+\varepsilon) \cap \mathcal{R}(5;l+1),$$

with $\varepsilon = 0$ or 1, there exists $m \in \mathcal{Q}_\varepsilon$ such that

$$n - m^4 \in [x, x+y] \cap \mathcal{R}(16;k) \cap \mathcal{R}(5;l).$$

Part (iii) of the lemma follows immediately.

Finally, we consider part (iv) of the lemma, and assume the hypotheses of case (iii). On noting that the conclusion of part (iv) is contained, in part, in case (iii) of the lemma, we see that it suffices to show that whenever

$$n \in [x+6561, x+6561+(y/40)^{4/3}] \cap \mathcal{R}(16;k+\varepsilon) \cap \mathcal{R}(5,l),$$

with $\varepsilon = 0$ or 1, then n is a B_{s+1}. But for such an integer n, on applying Lemma 12.3 with $\mathcal{H} = \mathcal{R}(2;\varepsilon) \cap \mathcal{R}(5;0)$ and $h = 10$, and observing that $m^4 \in \mathcal{R}(16;\varepsilon) \cap \mathcal{R}(5;0)$

whenever $m \in \mathcal{R}(2;\varepsilon) \cap \mathcal{R}(5;0)$, we deduce that there exists an integer $m \in \mathcal{R}(2;\varepsilon) \cap \mathcal{R}(5;0)$ such that
$$n - m^4 \in [x,\, x+y] \cap \mathcal{R}(16;\varepsilon) \cap \mathcal{R}(5;l).$$
The conclusion of part (iv) now follows. $\square$

We require one further ascent method that is a variant of the "U-type ascent" of Deshouillers and Dress [10]. This ascent gear makes fundamental use of the conclusion of the following lemma.

LEMMA 12.5. — *Let $n \in \mathcal{R}(16;1)$, and write*
$$\mathcal{T} = \{r \in \mathbb{Z} : r \equiv k \pmod{16} \text{ for some } k \text{ with } 4 \leqslant k \leqslant 12\}. \tag{12.25}$$
Then any set of five consecutive odd integers contains an element m satisfying the property that $(n - m^4)/16 \in \mathcal{T}$.

Proof. — Observe first that whenever a and b are integers satisfying $a - b \equiv \pm 7 \pmod{16}$, then either a or b belongs to $\mathcal{T}$. Next define the polynomial $f_n(l)$ by
$$f_n(l) = (n - (2l+1)^4)/16$$
$$= -l^4 - 2l^3 - \tfrac{1}{2}l(3l+1) + \tfrac{1}{16}(n-1).$$
Then the conclusion of the lemma follows on proving that for any set $\mathcal{L}$ of five consecutive integers, there exists an $l \in \mathcal{L}$ such that $f_n(l) \in \mathcal{T}$. But on observing that
$$f_n(a+2) - f_n(a) \equiv 8a^3 + 12a^2 + 2a - 7 \pmod{16},$$
one readily verifies that

when $u \equiv 0 \pmod 4$, one has $f_n(2u+2) - f_n(2u) \equiv -7 \pmod{16}$, (12.26)

when $u \equiv 1 \pmod 4$, one has $f_n(2u) - f_n(2u-2) \equiv -7 \pmod{16}$, (12.27)

when $u \equiv 2 \pmod 4$, one has $f_n(2u+3) - f_n(2u+1) \equiv 7 \pmod{16}$, (12.28)

when $u \equiv 3 \pmod 4$, one has $f_n(2u+1) - f_n(2u-1) \equiv 7 \pmod{16}$. (12.29)

Similarly, on noting that
$$f_n(a+1) - f_n(a) \equiv 12a^3 + 4a^2 + 3a + 11 \pmod{16},$$
one sees that

when $u \equiv 2 \pmod 8$, one has $f_n(2u+1) - f_n(2u) \equiv 7 \pmod{16}$, (12.30)

when $u \equiv 6 \pmod 8$, one has $f_n(2u-1) - f_n(2u-2) \equiv -7 \pmod{16}$, (12.31)

when $u \equiv 1 \pmod 8$, one has $f_n(2u+3) - f_n(2u+2) \equiv 7 \pmod{16}$, (12.32)

when $u \equiv 5 \pmod 8$, one has $f_n(2u+1) - f_n(2u) \equiv -7 \pmod{16}$. (12.33)

When $\mathcal{L}$ is a set of five consecutive integers, it takes the shape $\{2u-2, 2u-1, \ldots, 2u+2\}$, or else $\{2u-1, 2u, \ldots, 2u+3\}$. Thus, in view of the opening observation of this proof, the required conclusion follows in the former case from (12.26), (12.27) and (12.29)–(12.31), and in the latter case from (12.26), (12.28), (12.29), (12.32) and (12.33). In any case, therefore, the proof of the lemma is complete. □

LEMMA 12.6. — *Let x and y be real numbers with $x \geqslant 0$ and $y \geqslant 20000$, and let $\mathcal{T}$ be the set defined in (12.25). Suppose that every integer in the set $[x, x+y] \cap \mathcal{T}$ is a B_s. Then every integer in*

$$[16x + 6561, 16x + 6561 + (2y/5)^{4/3}] \cap \mathcal{R}(16;1)$$

is a B_{s+1}.

Proof. — When

$$n \in [16x + 6561, 16x + 6561 + (2y/5)^{4/3}] \cap \mathcal{R}(16;1),$$

we denote by $\mathcal{H}_n$ the set of all positive odd integers m satisfying the property that $(n - m^4)/16 \in \mathcal{T}$. Then by virtue of Lemma 12.5, we may apply Lemma 12.3 with $\mathcal{H} = \mathcal{H}_n$ and $h = 10$ to infer that there exists an $m \in \mathcal{H}_n$ satisfying $n - m^4 \in [16x, 16(x+y)]$. Fixing any such choice of m, and writing $n' = (n - m^4)/16$, we therefore find that $n' \in [x, x+y] \cap \mathcal{T}$. But then n' is a B_s, by assumption, and so we may conclude that $n = m^4 + 2^4 n'$ is a B_{s+1}. □

Having equipped ourselves with the necessary ascent tools, we are able to derive a result concerning the representation of small integers that, in combination with Theorem 3, suffices to complete the proof of the desired conclusion that $g(4) = 19$.

THEOREM 4. — *Every natural number not exceeding 10^{147} is a B_{19}.*

Proof. — When h is a natural number, we define $\phi_h(y) = (y/(4h))^{4/3}$. Also, for a function $\phi(y)$, we adopt the convention of writing $\phi^1(y) = \phi(y)$, and define $\phi^j(y)$ for $j \geqslant 2$ by putting $\phi^j(y) = \phi(\phi^{j-1}(y))$. Finally, it is convenient to write

$$x_0 = 1143348 \quad \text{and} \quad y_0 = 600^4.$$

We observe first that Lemma 12.2 asserts that every integer in

$$[x_0, x_0 + y_0] \cap \mathcal{R}(16;4) \cap \mathcal{R}(5;3)$$

is a B_8. Starting from this observation, we apply Lemma 12.4 (iii) four times in succession, and then we apply Lemma 12.4 (iv) four times in succession. In this way we find that every integer in

$$[x_0 + 4(81 + 6561), x_0 + 4(81 + 6561) + \phi_{10}^4(\phi_4^4(y_0))] \cap \mathcal{T}$$

is a B_{16}, where $\mathcal{T}$ is the set defined in (12.25). It therefore follows via three successive applications of Lemma 12.4 (ii) that every integer in

$$[x_0 + 26571, x_0 + 26571 + \phi_2^3(\phi_{10}^4(\phi_4^4(y_0)))] \cap \left(\bigcup_{k=4}^{15} \mathcal{R}(16;k)\right)$$

is a B_{19}. Meanwhile, one may deduce from Lemma 12.6 that every integer in
$$\left[16x_0 + 431649,\ 16x_0 + 431649 + \phi_{10}\big(16\phi_{10}^4(\phi_4^4(y_0))\big)\right] \cap \mathcal{R}(16;1)$$
is a B_{17}, whence by applying Lemma 12.4 (ii) twice, we find that every integer in
$$\left[16x_0 + 431651,\ 16x_0 + 431651 + \phi_2^2\big(\phi_{10}\big(16\phi_{10}^4(\phi_4^4(y_0))\big)\big)\right] \cap \left(\bigcup_{k=1}^{3} \mathcal{R}(16;k)\right)$$
is a B_{19}. Since a modicum of computation provides the estimates
$$\phi_2^3\big(\phi_{10}^4(\phi_4^4(y_0))\big) > 2 \times 10^{147}, \qquad \phi_2^2\big(\phi_{10}\big(16\phi_{10}^4(\phi_4^4(y_0))\big)\big) > 3 \times 10^{148},$$
and $16x_0 + 431651 < 2 \times 10^7$, we may conclude thus far that every integer in the set $[2 \times 10^7,\ 10^{147}] \setminus \mathcal{R}(16;0)$ is a B_{19}.

On the other hand, since $\phi_1^3(50000 - 13793) > 3 \times 10^7$, we find from the second conclusion of Lemma 12.1 (i), via three applications of Lemma 12.4 (i), that every integer in $[13793, 3 \times 10^7]$ is a B_{19}. Combining this conclusion with that of the previous paragraph and the first assertion of Lemma 12.1 (i), we at last conclude that every integer in $[1, 10^{147}] \setminus \mathcal{R}(16;0)$ is a B_{19}. But if n is a B_{19}, then so is $16^\nu n$ for each natural number ν. Then every integer in $[1, 10^{147}]$ is a B_{19}, and this completes the proof of the theorem. $\square$

BIBLIOGRAPHY

[1] T.M. APOSTOL – *Introduction to Analytic Number Theory*, Springer-Verlag, New York, 1976.

[2] R. BALASUBRAMANIAN – On Waring's problem: $g(4) \leqslant 20$, *Hardy-Ramanujan J.* **8** (1985), p. 1–40.

[3] R. BALASUBRAMANIAN, J.-M. DESHOUILLERS & F. DRESS – Problème de Waring pour les bicarrés, 1: Schéma de la solution, *C. R. Acad. Sci. Paris Sér. I Math.* **303** (1986), p. 85–88.

[4] ———, Problème de Waring pour les bicarrés, 2: résultats auxiliaires pour le théorème asymptotique, *C. R. Acad. Sci. Paris Sér. I Math.* **303** (1986), p. 161–163.

[5] H. DAVENPORT – On Waring's problem for fourth powers, *Ann. of Math.* **40** (1939), p. 731–747.

[6] J.-M. DESHOUILLERS – Le problème de Waring pour les bicarrés, in *Sém. Th. Nb. Bordeaux*, 1984/85, exp. 14.

[7] ———, Sur la majoration de sommes de Weyl biquadratiques, *Ann. Scuola Norm. Sup. Pisa Cl. Sci. (4)* **19** (1992), p. 291–304.

[8] J.-M. DESHOUILLERS & F. DRESS – Sommes de diviseurs et structures multiplicatives des entiers, *Acta Arith.* **49** (1988), p. 341–375.

[9] ———, Sums of 19 biquadrates: On the representation of large integers, *Ann. Scuola Norm. Sup. Pisa Cl. Sci. (4)* **19** (1992), p. 113–153.

[10] ———, Numerical results for sums of five and seven biquadrates and consequences for sums of 19 biquadrates, *Math. Comp.* **61** (1993), p. 195–207.

[11] J.-M. DESHOUILLERS, F. HENNECART, K. KAWADA, B. LANDREAU & T.D. WOOLEY – Survey (in preparation).

[12] J.-M. DESHOUILLERS, F. HENNECART & B. LANDREAU – Waring's problem for sixteen biquadrates – Numerical results, *J. Théor. Nombres Bordeaux* **12** (2000), p. 411–422.

[13] G.H. HARDY & J.E. LITTLEWOOD – Some problems in "Partitio Numerorum" (VI): Further researches on Waring's problem, *Math. Z.* **23** (1925), p. 1–37.

[14] A.E. INGHAM – *The Distribution of Prime Numbers*, Cambridge University Press, Cambridge, 1992.

[15] K. KAWADA & T.D. WOOLEY – Sums of fourth powers and related topics, *J. reine angew. Math.* **512** (1999), p. 173–223.

[16] B. LANDREAU – Moyennes de fonctions arithmétiques sur des suites de faible densité, Thèse 3e cycle, Université Bordeaux I, 1987.

[17] K.S. MCCURLEY – Explicit estimates for $\theta(x;3,l)$ and $\psi(x;3,l)$, *Math. Comp.* **42** (1984), p. 287–296.

[18] V.I. NEČAEV & V.L. TOPUNOV – Estimation of the modulus of complete rational trigonometric sums of degree three and four, *Trudy Mat. Inst. Steklov.* **158** (1981), p. 125–129, 229, Russian.

[19] J.B. ROSSER & L. SCHOENFELD – Approximate formulas for some functions of prime numbers, *Illinois J. Math.* **6** (1962), p. 64–94.

[20] H.E. THOMAS, JR. – A numerical approach to Waring's problem for fourth powers, Ph.D. Thesis, University of Michigan, 1973.

[21] R.C. VAUGHAN – *The Hardy-Littlewood method*, 2nd ed., Cambridge Univ. Press, Cambridge, 1997.

MÉMOIRES DE LA SOCIÉTÉ MATHÉMATIQUE DE FRANCE
Nouvelle série

2005
100. J.-M. DESHOUILLERS, K. KAWADA, T.D. WOOLEY – *On Sums of Sixteen Biquadrates*

2004
99. V. PASKUNAS – *Coefficient systems and supersingular representations of* $\mathrm{GL}_2(F)$
98. F.-X. DEHON – *Cobordisme complexe des espaces profinis et foncteur T de Lannes*
97. G.-V. NGUYEN-CHU – *Intégrales orbitales unipotentes stables et leurs transformées de Satake*
96. J.-L. WALDSPURGER – *Une conjecture de Lusztig pour les groupes classiques*

2003
95. T. ROBLIN – *Ergodicité et équidistribution en courbure négative*
94. P.T. CHRUŚCIEL, E. DELAY – *On mapping properties of the general relativistic constraints operator in weighted function spaces, with applications*
93. F. BERNON – *Propriétés de l'intégrale de Cauchy Harish-Chandra pour certaines paires duales d'algèbres de Lie*
92. C. SABOT – *Spectral properties of self-similar lattices and iteration of rational maps*

2002
91. J.-M. DELORT – *Global solutions for small nonlinear long range perturbations of two dimensional Schrödinger equations*
90. P.-Y. JEANNE – *Optique géométrique pour des systèmes semi-linéaires avec invariance de jauge*
89. F. PIERROT – K-*théorie et multiplicités dans* $L^2(G/\Gamma)$
88. O. LAFITTE – *The wave diffracted by a wedge with mixed boundary conditions*

2001
87. L. BARBIERI-VIALE, V. SRINIVAS – *Albanese and Picard* 1-*motives*
86. E. BUJALANCE, F.-J. CIRRE, J.-M. GAMBOA, G. GROMADZKI – *Symmetry types of hyperelliptic Riemann surfaces*
85. M. HARRIS, S. ZUCKER – *Boundary cohomology of Shimura varieties, III : Coherent cohomology on higher-rank boundary strata and applications to Hodge theory*
84. B. PERRIN-RIOU – *Théorie d'Iwasawa des représentations* p-*adiques semi-stables*

2000
83. J. SJÖSTRAND – *Complete asymptotics for correlations of Laplace integrals in the semi-classical limit*
82. S. RIGOT – *Ensembles quasi-minimaux avec contrainte de volume et rectifiabilité uniforme*
81. P. BERTHELOT – $\mathcal{D}$-*modules arithmétiques II. Descente par Frobenius*
80. V. MAILLOT – *Géométrie d'Arakelov des variétés toriques et fibrés en droites intégrables*

1999
79. P. LE CALVEZ – *Décomposition des difféomorphismes du tore en applications déviant la verticale (avec la collaboration de J.-M. GAMBAUDO)*
78. S. CHOI – *The Convex and Concave Decomposition of Manifolds with Real Projective Structures*
77. E. RISLER – *Linéarisation des perturbations holomorphes des rotations et applications*
76. J.-P. SCHNEIDERS – *Quasi-Abelian Categories and Sheaves*

1998
75. C. CHEVERRY – *Systèmes de lois de conservation et stabilité BV*
74. M.-C. ARNAUD – *Le « closing lemma » en topologie* C^1
72/73. J. WINKELMANN – *Complex Analytic Geometry of Complex Parallelizable Manifolds*

1997

71. K. THOMSEN – *Limits of certain subhomogeneous C^*-algebras*
70. B. LEMAIRE – *Intégrales orbitales sur $GL(N,F)$ où F est un corps local non archimédien*
69. F. COURTÈS – *Sur le transfert des intégrales orbitales pour les groupes linéaires (cas p-adique)*
68. E. LEICHTNAM, P. PIAZZA – *The b-pseudodifferential calculus on Galois coverings and a higher Atiyah-Patodi-Singer index theorem*

1996

67. H. HIDA – *On the search of genuine p-adic modular L-functions for $\mathrm{GL}(n)$*
66. F. LOESER – *Faisceaux pervers, transformation de Mellin et déterminants*
65. N. BARDY – *Systèmes de racines infinis*
64. M. KASHIWARA, P. SCHAPIRA – *Moderate and formal cohomology associated with constructible sheaves*

1995

63. M. KASHIWARA – *Algebraic Study of Systems of Partial Differential Equations (Master's Thesis, Tokyo University, December 1970)*
62. S. DAVID – *Minorations de formes linéaires de logarithmes elliptiques*
61. J.-P. LABESSE – *Noninvariant base change identities*
60. G. LEBEAU – *Propagation des ondes dans les dièdres*

1994

59. A. BOMMIER – *Prolongement méromorphe de la matrice de diffusion pour les problèmes à N corps à longue portée*
58. F. CHOUCROUN – *Analyse harmonique des groupes d'automorphismes d'arbres de Bruhat-Tits*
57. E. ANDRONIKOF – *Microlocalisation tempérée*
56. B. SÉVENNEC – *Géométrie des systèmes hyperboliques de lois de conservation*

1993

55. N. BURQ – *Contrôle de l'équation des plaques en présence d'obstacles strictement convexes*
54. L. RAMELLA – *Sur les schémas définissant les courbes rationnelles lisses de $\mathbf{P}^3$ ayant fibré normal et fibré tangent restreint fixés*
53. E. LEICHTNAM – *Le problème de Cauchy ramifié linéaire pour des données à singularités algébriques*
52. L. BLASCO – *Paires duales réductives en caractéristique 2*
    P.J. SALLY Jr., M. TADIC – *Induced representations and classification for $\mathrm{GSp}(2,F)$ and $\mathrm{Sp}(2,F)$*

1992

51. P. KERDELHUÉ – *Spectre de l'opérateur de Schrödinger magnétique avec symétrie d'ordre six*
50. A. ARRONDO, I. SOLS – *On congruences of lines in the projective space (Chapter 6 written in collaboration with M. Pedreira)*
49. A. AMBROSETTI – *Critical points and nonlinear variational problems — Cours de la chaire Lagrange*
48. M.-C. ARNAUD – *Type des points fixes des difféomorphismes symplectiques de $\mathbf{T}^n \times \mathbf{R}^n$*

CNRS EDITIONS

un éditeur au service de l'édition scientifique

Einstein et la relativité générale
Les chemins de l'espace-temps

Jean Eisenstaedt

Comment, dans quel contexte, et au prix de quel effort la théorie de la relativité a vu le jour et évolué ? Cet ouvrage de vulgarisation nous donne le fil conducteur de cette aventure et associe intimement l'histoire des sciences et l'aspect biographique, en citant des journaux ou des correspondances d'astronomes ou de physiciens proches d'Einstein, découragés, enthousiastes ou même agressifs face à cette théorie difficile à accepter, à comprendre.

L'auteur insiste en particulier sur la « traversée du désert » d'Einstein, et sur la difficile institutionnalisation de la théorie. Les structures de la recherche en relativité sont restées longtemps artisanales ; il n'y a pas eu d'enseignement suivi sur la relativité avant les années 1950.

L'élaboration de la théorie, replacée dans le contexte de l'époque, est pour ainsi dire vécue de l'intérieur par le lecteur qui en découvre le développement heurté, sa croissance lente et son douloureux manque de résultats face à la théorie quantique.

On y comprend notamment comment les trous noirs, qui n'ont pu être posés ni pensés lors de la naissance de la théorie, vont être « inventés », compris, acceptés dans les années 1970... permettant une interprétation révolutionnaire de la théorie qui conduira au renouveau actuel.

Collection CNRS Histoire des sciences
348 pages, 29 €

Pour trouver et commander nos ouvrages :

LA LIBRAIRIE de CNRS ÉDITIONS, 151 bis, rue Saint-Jacques - 75005 PARIS
Tél. : 01 53 10 05 05 - Télécopie : 01 53 10 05 07 - Mél : librairie@cnrseditions.fr

Site Internet : www.cnrseditions.fr
Frais de port par ouvrage : France : 5 € - Etranger : 5,5 €

Pour plus de renseignements, n'hésitez pas à contacter
le Service clientèle de CNRS ÉDITIONS, 15, rue Malebranche - 75005 Paris
Tél : 01 53 10 27 07/09 - Télécopie : 01 53 10 27 27 - Mél : cnrseditions@cnrseditions.fr

Nouveautés
Panoramas et Synthèses

RATIONAL REPRESENTATIONS, THE STEENROD ALGEBRA AND FUNCTOR HOMOLOGY

Vincent Franjou, Eric M. Friedlander, Teimuraz Pirashvili, Lionel Schwartz

Ce livre traite d'algèbre homologique dans les catégories de foncteurs, avec une attention particulière pour les foncteurs polynomiaux entre espaces vectoriels sur un corps fini. Il en présente des applications dans trois domaines des mathématiques : la théorie des représentations, la topologie algébrique et la K-théorie. A chacune de ces applications, les catégories de foncteurs apportent des avancées théoriques et des outils de calcul puissants. D'abord, T. Pirashvili expose les bases de la théorie. E. Friedlander l'applique alors aux représentations rationnelles des groupes linéaires. L. Schwartz établit les relations de l'algèbre de Steenrod avec les catégories de foncteurs. Enfin, V. Franjou et T. Pirashvili présentent un théorème de Scorichenko : la K-théorie stable est l'homologie des foncteurs.

La théorie des motifs, introduite par A. Grothendieck il y a 40 ans et demeurée longtemps conjecturale, a connu depuis une quinzaine d'années des développements spectaculaires. Ce texte a pour objectif de rendre ces avancées accessibles au non-spécialiste, tout en donnant, au cours de ses deux premières parties, une vision unitaire des fondements géométriques de la théorie (pure et mixte). La troisième partie, consacrée aux périodes des motifs, en propose une illustration concrète ; on y traite en détail les exemples des valeurs de la fonction gamma aux points rationnels, et des nombres polyzêta.

UNE INTRODUCTION AUX MOTIFS
(MOTIFS PURS, MOTIFS MIXTES, PÉRIODES)

Yves André

Prix public* : 26 € ; Prix membre* : 19 €
* Frais de port non compris

Commandes
Maison de la SMF, BP 67, 13274 Marseille Cedex 9 France
Tél : 04 91 26 74 64 - Fax : 04 91 41 17 51 - mail : smf@smf.univ-mrs.fr
url : http://smf.emath.fr/